Student's Choice

Regents Review

Integrated Algebra

Henry Gu

Mathematics Teacher
John Dewey High School
Brooklyn, New York

Disclaimer: The contents of this book are the author's alone and not those of the New York City Department of Education.

Author: Henry Gu
Editor: Christopher Gu

www.hsmathreview.com

ISBN-10: 1453880984   ISBN-13: 9781453880982

"Everything should be made as simple as possible, but not simpler."

\- Albert Einstein

Preface

"Different books, different results." This book is different from the lengthy review books. It is designed to help students review all the important math topics when they have only six to eight weeks before the Regents exam. This book uses real Regents questions and shows all the necessary steps to solve these problems. Its clear format is like no other.

This book is structured in three parts:
1. An Algebra review that will help students remember all the key topics and build their problem solving skills through the use of examples.
2. A practice section with real Regents questions.
3. Answers and explanations.

The topics for the practice questions correspond to the sections in the Algebra review. Students can easily refer back to the matching review sections, while they are doing the practice.

This review book is geared towards helping students succeed with high scores on the Regents exams. I have already used these review sheets with my own Regents classes and I have seen firsthand that their performance is significantly higher than the statewide average. Both students and teachers like these review sheets because they are straightforward and practical.

For updates and information on additional titles, please visit our website www.hsmathreview.com.

Acknowledgement

Thanks to the teachers and students at John Dewey High School who have already used these review sheets for their own Regents review and have achieved excellent scores.

Thanks to my family for their unconditional love and support.

Dedication

This book is dedicated to all the students taking the Regents exams. I wish you the best of luck!

Contents

Contents

PART 1. Algebra

I. SETS, NUMBERS, AND OPERATIONS

1. Set and Notation

A set is a collection of distinct elements.

Set Notation:
Finite Sets:
  e.g. { 1, 2, 3, 4, 5 } , { a, b, c }
Infinite Sets:
  e.g. { 1, 2 , 3, 4, 5 ••• } ,
  $\{\frac{1}{2}, \frac{1}{4}, \frac{1}{8}, \frac{1}{16} ••• \}$
Empty Set or Null Set:
    { } or Ø
  e.g. {0} is not an empty set.

Set Builder Notation:
  e.g. { x | 0 ≤ x ≤ 10, where x is a real number }

Interval Notation:
  e.g. (2, 5) represents { x | 2 < x < 5 }
        [2, 5] represents { x | 2 ≤ x ≤ 5 }
        (2, 5] represents { x | 2 < x ≤ 5 }
        [2, 5) represents { x | 2 ≤ x < 5 }

e.g. Which of the following notation is equivalent
  to the set {1, 2, 3, 4} ?
(1). { x | 1 < x < 4, where x is a whole number }
(2). { x | 0 < x < 4, where x is a whole number }
(3). { x | 1 < x ≤ 4, where x is a whole number }
(4). { x | 0 < x ≤ 4, where x is a whole number }
Here (1) is { 2, 3 };
    (2) is { 1, 2, 3 };
    (3) is { 2, 3, 4 };
    (4) is { 1, 2, 3, 4 }
Answer: (4)

2. Operations with Sets

Universe or Universal Set:
The set of all elements under consideration.

Subset:
A set which is a part of a larger set.
e.g. { 1, 2, 3 } is a subset of { 1, 2, 3, 4, 5 }

Complement of a set:
e.g. Universal Set U = { 1, 2, 3, 4, 5, 6, 7, 8 }
      Set A = { 1, 2, 3 }
The complement of set A is denoted by $\overline{A}$ or A'.
      $\overline{A}$ = { 4, 5, 6, 7, 8 }
The complement of set A has all elements in the
universal set except the elements in set A.

Intersection of two sets: symbol ∩
The set of all elements that belong to both sets.

Union of two sets: symbol ∪
The set of all elements in either set.

e.g. A = { 1, 2, 3, 4, 5 } and B = { 2, 4, 6, 8, 10 }

    then A ∩ B = { 2, 4 }
          A ∪ B = { 1, 2, 3, 4, 5, 6, 8, 10 }

3. Numbers

Counting Numbers or Natural Numbers:
      1, 2, 3, 4, 5, ...
Whole Numbers: 0, 1, 2, 3, 4, 5, ...
Integers: ..., -3, -2, -1, 0, 1, 2, 3, ...
Consecutive Integers: n, n+1, n+2, ...
    e.g. -3, -2, -1 ;   4, 5, 6
Consecutive Even Integers: n, n+2, n+4, ...
    e.g. -8, -6, -4 ;   0, 2, 4   (zero is an even number)
Consecutive Odd Integers: n, n+2, n+4, ...
    e.g. -5, -3, -1 ;   7, 9, 11
Perfect Squares: 4, 9, 16, 25, 36, 49, 64, 81, ...
Prime Numbers: 2, 3, 5, 7, 11, 13, 17, 19, 23, ...
Rational Numbers can be written as an integer, a quotient of
two integers, a terminating or repeating decimal:
    e.g.   2, -5, 1.25, 0.333..., 2.345345...,
        $\sqrt{4} = 2$, $\frac{4}{5}$, $\sqrt{\frac{4}{9}} = \frac{2}{3}$
Irrational Numbers are decimal numbers that neither repeat
nor terminate:
          $\sqrt{3}$, 1.41421... , π
Real Numbers: All the above numbers.
Rounding: 3.456 rounded to the nearest integer is 3, to the
nearest tenth is 3.5, and to the nearest hundredth is 3.46
Absolute Value of a Number: $|5| = 5$, $|-5| = 5$,
        $|12| - |-5| = 12 - 5 = 7$

  e.g. The integers are a subset of the rational numbers.
        The rational numbers are a subset of the real numbers.
        The unoin of the rational numbers and the irrational
        numbers is the set of real numbers.

4. Properties of Operations

Commutative: $a + b = b + a$, $ab = ba$
$$3 + 5 = 5 + 3 \;, \;\; 3 \bullet 5 = 5 \bullet 3$$
Associative: $a + (b + c) = (a + b) + c$
$$a \bullet (b \bullet c) = (a \bullet b) \bullet c$$
$$3 + (5 + 7) = (3 + 5) + 7, \;\; 3(5 \bullet 7) = (3 \bullet 5)7$$
Distributive: $a(b + c) = ab + ac$
$$3(5 + 7) = 3 \bullet 5 + 3 \bullet 7$$
Additive Identity: 0
$$x + 0 = x, \;\;\; 0 + x = x$$
$$5 + 0 = 5, \;\;\; 0 + 5 = 5$$
Multiplicative Identity: 1
$$x \bullet 1 = x, \;\;\; 1 \bullet x = x$$
$$3 \bullet 1 = 3, \;\;\; 1 \bullet 3 = 3$$
Additive Inverse: -x
$$x + (-x) = 0 \;\;\;\;\; -5 + 5 = 0$$
Multiplicative Inverse: $\dfrac{1}{x}$

$$x \bullet \dfrac{1}{x} = 1 \;\;\;\;\;\;\; 3 \bullet \dfrac{1}{3} = 1$$
e.g. If $\odot$ is a binary operation defined by
$a \odot b = a^2 + b^2$, then $3 \odot 4 = 3^2 + 4^2 = 25$.

5. Ratio, Proportion, and Percentage

If two ratios are equal, they are in proportion.
$$\dfrac{a}{b} = \dfrac{c}{d} \;\;\; \text{or} \;\;\; a \bullet d = b \bullet c$$
In a proportion, the product of the means is equal to the product of the extremes.
e.g. The ratio of the three interior angles of a triangle
is 2 : 3 : 4. Find the measure of the largest angle.
$$2x + 3x + 4x = 180$$
$$9x = 180$$
$$x = 20$$
The measure of the largest angle is
$$4x = 80$$

Percent:

$$\% = \dfrac{1}{100} \;, \;\;\; 100\% = 1$$

$$60\% = \dfrac{60}{100} = 0.6, \;\;\; 6\% = \dfrac{6}{100} = 0.06$$

$0.6 = 0.6 \times 100\% = 60\%$
6% of $\$50 = 0.06 \bullet \$50 = \$3$

Percent of Increase or Decrease:

$$\dfrac{|\text{New Amount - Original Amount}|}{\text{Original Amount}} \bullet 100\%$$

e.g. The gas price increased to \$2.50/gal from \$2.00/gal.

$$\text{Percent of Increase} = \dfrac{|2.50 - 2.00|}{2.00} \bullet 100\% = 25\%$$

Tax Problems:

Tax = Base Price • Tax Rate
e.g. The tax rate in NYC is 8.5%. How much do you pay
for a merchandise tagged \$50?
Tax = $50 \bullet 8.5\% = 50 \bullet 0.085 = \$4.25$
Total Amount = $50 + 4.25 = \$54.25$

II. ALGEBRAIC EXPRESSIONS AND OPERATIONS

1. Operations of Polynomials

Combine Like Terms:
$$\begin{aligned} \text{e.g.} \quad & 3x^2 + x - 8 - x^2 + 5x + 4 \\ = & (3x^2 - x^2) + (x + 5x) + (-8 + 4) \\ = & 2x^2 + 6x - 4 \end{aligned}$$

Multiply:
$$\begin{aligned} \text{e.g.} \quad & 3y(2x^2 + 2y^2 - 2) \\ = & 3y \bullet 2x^2 + 3y \bullet 2y^2 + 3y(-2) \\ = & 6x^2y + 6y^3 - 6y \end{aligned}$$

$$\begin{aligned} \text{e.g.} \quad & (a + b)(a + b) \\ = & a \bullet a + a \bullet b + b \bullet a + b \bullet b \\ = & a^2 + 2ab + b^2 \end{aligned}$$

Divide:
$$\dfrac{12x^4 - 4x^2}{4x^2}$$
$$= \dfrac{12x^4}{4x^2} - \dfrac{4x^2}{4x^2}$$
$$= 3x^2 - 1$$

2. Factoring Polynomials

Greatest Common Factor (GCF):
$$3x^2 + 6x = 3x(x + 2)$$
$$2y^3 - 4y^2 + 2y = 2y(y^2 - 2y + 1)$$

The Difference of Two Squares:
$$a^2 - b^2 = (a + b)(a - b)$$
$$4y^2 - 25 = (2y + 5)(2y - 5)$$

Trinomial:
$$x^2 + 2x - 15 = (x + 5)(x - 3)$$
Here $5 \cdot (-3) = -15$ and $5 + (-3) = 2$

Factor Completely:
$$2x^3 - 14x^2 + 20x$$
$$= 2x(x^2 - 7x + 10)$$
$$= 2x(x - 2)(x - 5)$$

3. Rational Expressions

Denominator cannot be zero.

e.g. $\dfrac{x}{x-2}$ when $x = 2$, $x - 2 = 0$, undefined

Simplify: $\dfrac{2x^3}{x^2 - x - 12} \cdot \dfrac{x^2 - 16}{6x}$

$= \dfrac{2x^3(x+4)(x-4)}{(x+3)(x-4) \cdot 6x}$ factor the numerator

and the denominator first;

$= \dfrac{x^2(x+4)}{3(x+3)}$ cancel out common factors in

the numerator and the denominator

Divide: $\dfrac{\dfrac{x-3}{2x+1}}{\dfrac{2x-6}{2x}}$

$= \dfrac{x-3}{2x+1} \cdot \dfrac{2x}{2x-6}$ multiply inverse

$= \dfrac{(x-3) \cdot 2x}{(2x+1) \cdot 2(x-3)}$

$= \dfrac{x}{(2x+1)}$

Combine: $\dfrac{1}{x+1} + \dfrac{x-1}{2}$ LCD $= 2(x+1)$

$= \dfrac{2 \cdot 1}{2(x+1)} + \dfrac{(x-1) \cdot (x+1)}{2(x+1)}$

$= \dfrac{2 + x^2 - 1}{2(x+1)}$

$= \dfrac{x^2 + 1}{2(x+1)}$

4. Radicals

$$\sqrt{a \cdot b} = \sqrt{a} \cdot \sqrt{b} \qquad a \geq 0, b \geq 0$$
$$\sqrt{\dfrac{a}{b}} = \dfrac{\sqrt{a}}{\sqrt{b}} \qquad a \geq 0, b > 0$$

Simplify: $\sqrt{75} = \sqrt{25 \cdot 3} = 5\sqrt{3}$
$\sqrt{300} = \sqrt{100 \cdot 3} = 10\sqrt{3}$

Combine: $5\sqrt{x} + 3\sqrt{x} = 8\sqrt{x}$
$\sqrt{18} - 4\sqrt{2} - \sqrt{9 \cdot 2} - 4\sqrt{2} = 3\sqrt{2} - 4\sqrt{2} = -\sqrt{2}$

Multiply: $2\sqrt{3} \cdot 4\sqrt{5} = 2 \cdot 4\sqrt{3 \cdot 5} = 8\sqrt{15}$
$3\sqrt{2} \cdot 7\sqrt{2} = 3 \cdot 7\sqrt{2 \cdot 2} = 21 \cdot 2 = 42$

Divide: $\dfrac{4\sqrt{6}}{2\sqrt{3}} = \dfrac{4}{2}\sqrt{\dfrac{6}{3}} = 2\sqrt{2}$

Rationalize: $\dfrac{1}{\sqrt{3}} = \dfrac{1}{\sqrt{3}} \cdot \dfrac{\sqrt{3}}{\sqrt{3}} = \dfrac{\sqrt{3}}{3}$

5. Exponents

$a^0 = 1$ $(a \neq 0)$ $5^0 = 1$, $(-5)^0 = 1$, $1.2^0 = 1$

$x^{-n} = \dfrac{1}{x^n}$ $(x \neq 0)$ $5^{-2} = \dfrac{1}{5^2} = \dfrac{1}{25}$

$x^a \cdot x^b = x^{a+b}$ $5^2 \cdot 5^3 = 5^{2+3} = 5^5$

$\dfrac{x^a}{x^b} = x^{a-b}$ $\dfrac{8xy^3}{2xy} = \dfrac{8}{2} \cdot \dfrac{x}{x} \cdot \dfrac{y^3}{y} = 4y^2$

$(x^a)^b = x^{a \cdot b}$ $(5^2)^3 = 5^{2 \cdot 3} = 5^6$

e.g. $(-5)^2 = (-5)(-5) = 25$
$-5^2 = -(5^2) = -25$

Scientific Notation:

$a \times 10^n$ $1 \leq a < 10$, n an integer

e.g. $23,000 = 2.3 \times 10^4$
$0.0043 = 4.3 \times 10^{-3}$

6. Evaluation of Algebraic Expressions and Formulas

e.g. If $x = 4$, $y = -3$ then
$x^2 - 4y = (4)^2 - 4(-3) = 16 + 12 = 28$

e.g. Solve for L in terms of P and W
$P = 2L + 2W$
$P - 2W = 2L$
$\dfrac{P - 2W}{2} = L$ $L = \dfrac{P - 2W}{2}$

III. EQUATIONS AND INEQUALITIES

1. Linear Equations

$4(x + 1) = 2x + 10$ remove the ( ) first
$4x + 4 = 2x + 10$ combine variables on one side,
            numbers on the other side;
$4x - 2x = 10 - 4$ change the term's sign when
            across the $=$ sign
$2x = 6$
$x = 3$

e.g.   Solve for x: $2ax - 5x = 2$
            $(2a - 5)x = 2$
            $x = \dfrac{2}{2a - 5}$

e.g. $\dfrac{2}{3}x - 6 = \dfrac{1}{2}x + 4$
$4x - 36 = 3x + 24$ multiply both sides by LCD $= 6$
$4x - 3x = 24 + 36$
$x = 60$

Verbal Problems

Average Speed: $s = \dfrac{d}{t}$ d: Total distance, t: time

e.g. A car travels 300 miles in 5 hours, the average speed
$s = \dfrac{300}{5} = 60$ miles/hr.

e.g. Tom drove 120 miles to his friend's house and the same distance back home. It took him 2 hours to drive there and 3 hours to drive back.
The average speed of driving out:
$s_1 = \dfrac{120}{2} = 60$ miles/hr.
The average speed of driving back:
$s_2 = \dfrac{120}{3} = 40$ miles/hr.
The average speed of the whole trip:
$s = \dfrac{120 + 120}{3 + 2} = \dfrac{240}{5} = 48$ miles/hr.
(Note: $s \neq \dfrac{s_1 + s_2}{2} = 50$ miles/hr.)

2. Linear Inequalities

Solving a linear inequality is the same as solving a linear equation except when both sides of the inequality are multiplied or divided by a negative number, the inequality sign is reversed.

e.g. $3x - 10 \geq 2$
        $3x \geq 2 + 10$
        $3x \geq 12$
        $x \geq 4$

e.g. $3 - 2x > 9$
        $-2x > 9 - 3$
        $\dfrac{-2x}{-2} < \dfrac{6}{-2}$ inequality sign reversed
        $x < -3$

3. Quadratic Equations

e.g. $x^2 - 10 = 3x$
        $x^2 - 3x - 10 = 0$ set one side equal to zero
        $(x + 2)(x - 5) = 0$ factor the trinomial
        $x + 2 = 0$ or $x - 5 = 0$
        $x = -2$ or $x = 5$
        solution set $\{ -2, 5\}$

e.g. $2x^2 = 5x$
        $2x^2 - 5x = 0$
        $x(2x - 5) = 0$
        $x = 0$ or $2x - 5 = 0$
                $2x = 5$
                $x = \dfrac{5}{2}$
        solution set $\{ 0 , \dfrac{5}{2} \}$

e.g. $x^2 + 5 = 30$
        $x^2 + 5 - 30 = 0$
        $x^2 - 25 = 0$
        $(x + 5)(x - 5) = 0$
        $x + 5 = 0$ or $x - 5 = 0$
        $x = -5$ or $x = 5$

4. Rational Equations

Use Cross-Multiplying Method:

e.g. $\dfrac{x+2}{x-3} = \dfrac{3}{4}$

$4(x+2) = 3(x-3)$ cross-multiply

$4x + 8 = 3x - 9$

$4x - 3x = -9 - 8$

$x = -17$ check   (denominator can not be zero)

Use LCD Method:

e.g. $\dfrac{2}{x-1} + \dfrac{1}{2} = \dfrac{4}{x-1}$ LCD $= 2(x-1)$

$4 + (x-1) = 8$ multiply LCD on both sides

$x - 1 = 4$

$x = 5$ check: True

5. Linear System

Substitution Method:

$2x + y = 6$ (1)

$x = 3y + 10$ (2)

Substitute x by $3y + 10$ in Eq.(1):

$2(3y + 10) + y = 6$

$6y + 20 + y = 6$

$6y + y = 6 - 20$

$7y = -14$

$y = -2$

Use Eq.(2) $x = 3(-2) + 10 = -6 + 10 = 4$

Solution: $x = 4$, $y = -2$ or $(4, -2)$

Addition or Subtration Method:

$x + y = 7$ (1)

$3x - 2y = 1$ (2)

Multiply Eq.(1) by 2:

$2x + 2y = 14$ (3)

Add Eq.(3) and Eq.(2)

$\begin{array}{r} 2x + 2y = 14 \\ 3x - 2y = 1 \\ \hline 5x = 15 \end{array}$

$x = 3$

Substitute x by 3 in Eq. (1):

$3 + y = 7$

$y = 4$

Solution: $x = 3$, $y = 4$ or $(3, 4)$

e.g.

3 slices of pizza and 2 colas cost $6.00.

2 slices of pizza and 3 colas cost $5.25.

Find the price of one slice of pizza and one cola.

p: price of one pizza

c: price of one cola

$3p + 2c = 6$ (1)

$2p + 3c = 5.25$ (2)

Eq.(1) x 3

$9p + 6c = 18$ (3)

Eq.(2) x 2

$4p + 6c = 10.5$ (4)

Eq.(3) - Eq.(4)

$5p = 7.5$

$p = 1.5$

Replace p by 1.5 in Eq.(1)

$3(1.5) + 2c = 6$

$4.5 + 2c = 6$

$2c = 1.5$

$c = 0.75$

Answer: one slice of pizza is $1.50.

one cola is $0.75.

6. Quadratic-Linear System

e.g. $y = x^2 - 8$ (1)

$y + 5 = 2x$ (2)

From Eq. (2) $y = 2x - 5$ (3)

Substitute y by $2x - 5$ in Eq.(1):

$2x - 5 = x^2 - 8$

$x^2 - 2x - 3 = 0$

$(x - 3)(x + 1) = 0$

$\begin{array}{l|l} x - 3 = 0 & x + 1 = 0 \\ x = 3 & x = -1 \\ y = 2(3) - 5 = 1 & y = 2(-1) - 5 = -7 \end{array}$

Solution: $\{(3, 1) , (-1, -7)\}$

IV. RIGHT TRIANGLES AND TRIGONOMETRY

1. Pythagorean Theorem

$$a^2 + b^2 = c^2$$

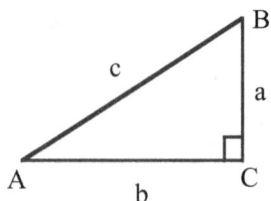

∠C is a right angle
a and b are legs
c is the hypotenuse

Pythagorean Triples:
    3, 4, 5;   6, 8, 10;   9, 12, 15 etc.
    5, 12, 13;   10, 24, 26 etc.

e.g. The ratio of two legs are 3:4 and the hypotenuse is 15.

Find the lengths of the two legs:
$$(3n)^2 + (4n)^2 = 15^2$$
$$9n^2 + 16n^2 = 15^2$$
$$25n^2 = 225$$
$$n^2 = 9$$
    $n = 3$ ($n = -3$ rejected)
$3n = 3 \bullet 3 = 9$ and $4n = 4 \bullet 3 = 12$
The lengths of the two legs are 9 and 12.

2. Trigonometric Ratios

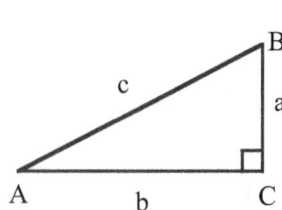

$$\sin A = \frac{\text{Opp}}{\text{Hyp}} = \frac{a}{c}$$

$$\cos A = \frac{\text{Adj}}{\text{Hyp}} = \frac{b}{c}$$

$$\tan A = \frac{\text{Opp}}{\text{Adj}} = \frac{a}{b}$$

3. Applications

Determine the right triangle and use the trigonometric ratios to solve the problem.

e.g.

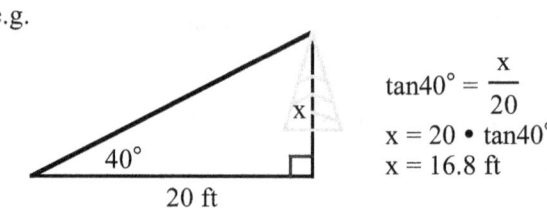

$$\tan 40° = \frac{x}{20}$$
$$x = 20 \bullet \tan 40°$$
$$x = 16.8 \text{ ft}$$

e.g.
A ladder is leaning against a vertical wall, making an angle of 60° with the ground and reaching a height of 12 feet on the wall.
Find, to the nearest foot, the length of the ladder.
Find, to the nearest foot, the distance from the base of the ladder to the wall.

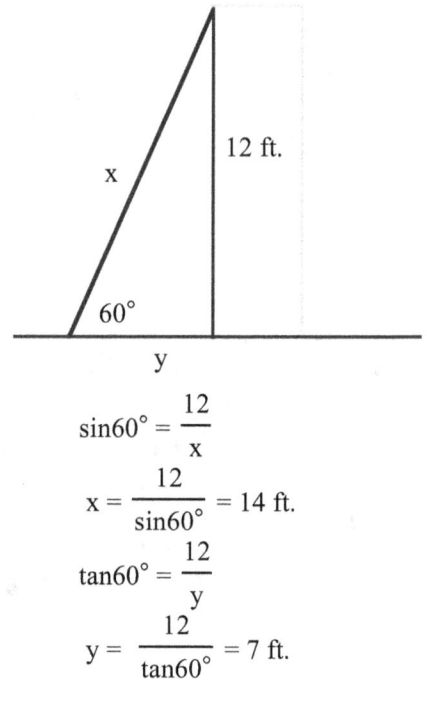

$$\sin 60° = \frac{12}{x}$$

$$x = \frac{12}{\sin 60°} = 14 \text{ ft.}$$

$$\tan 60° = \frac{12}{y}$$

$$y = \frac{12}{\tan 60°} = 7 \text{ ft.}$$

e.g.   angle of depression

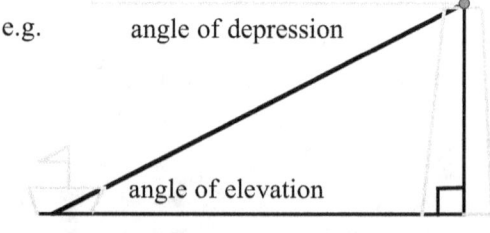

angle of elevation

angle of elevation = angle of depression

PART 2. Geometry and Measurements

V. BASIC GEOMETRY

1. Angles

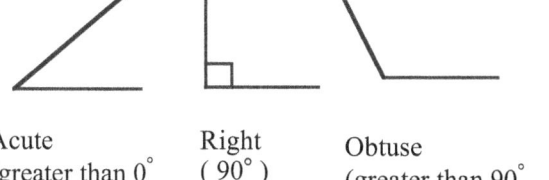

Acute   Right
(greater than $0°$ ($90°$)   Obtuse
and less than $90°$)   (greater than $90°$
                                and less than $180°$)

If $\angle A$ and $\angle B$ are complementary, then
$\qquad m\angle A + m\angle B = 90$ vice versa.
If $\angle A$ and $\angle B$ are supplementary, then
$\qquad m\angle A + m\angle B = 180$ vice versa.
A linear pair of angles are supplementary.
Vertical angles are congruent.

e.g.

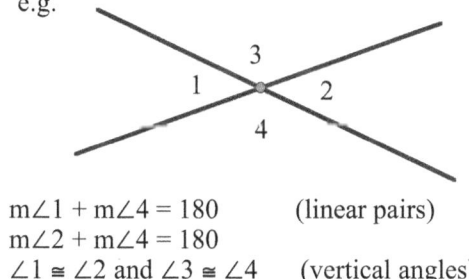

$m\angle 1 + m\angle 4 = 180$ (linear pairs)
$m\angle 2 + m\angle 4 = 180$
$\angle 1 \cong \angle 2$ and $\angle 3 \cong \angle 4$ (vertical angles)

2. Parallel Lines

Parallel lines are everywhere equidistant.

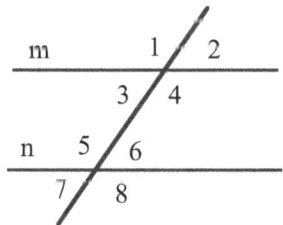

If line m ‖ line n, then
$\angle 3 \cong \angle 6$, $\angle 4 \cong \angle 5$ (alternate interior angles)
$\angle 1 \cong \angle 5$, $\angle 2 \cong \angle 6$ (corresponding angles)
$\angle 3 \cong \angle 7$, $\angle 4 \cong \angle 8$
$m\angle 3 + m\angle 5 = 180$ (interior angles on the same
$m\angle 4 + m\angle 6 = 180$ side of the transversal)

3. Perpendicular Lines

Two lines are perpendicular if they form right angles.
  e.g. If $\overline{AB} \perp \overline{BC}$, then $m\angle ABC = 90$.

4. Triangles

The sum of the three interior angles = $180°$;
exterior angle = the sum of 2 nonadjacent interior angles.

e.g.

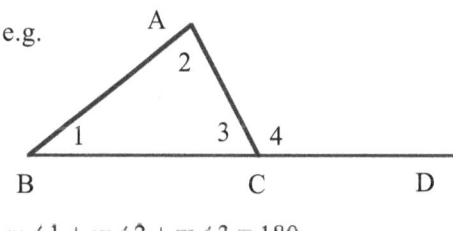

$m\angle 1 + m\angle 2 + m\angle 3 = 180$

$m\angle 1 + m\angle 2 = m\angle 4$

Any side is greater than the difference of the
other 2 sides and less than the sum of them.
$$\left|s_1 - s_2\right| < s_3 < \left|s_1 + s_2\right|$$

e.g. If two sides of a triangle are 3 and 5,
    then the 3rd side s_3 is
$$\left|3 - 5\right| < s_3 < \left|3 + 5\right|,$$
$$2 < s_3 < 8$$

Equilateral $\triangle$: three equal sides and three equal interior
angles ($60°$ each).
Isosceles $\triangle$: two equal sides and two equal base angles.

5. Quadrilaterals

Quadrilateral: 4-sided polygons.
Parallelogram: Opposite sides are parallel;
                Opposite sides are congruent;
                Opposite angles are congruent;
                Adjacent angles are supplementary;
                Diagonals bisect each other.
Rhombus: All the properties of the parallelogram;
           4 sides are congruent;
           Diagonals are perpendicular;
           Diagonals bisect the interior angles.
Rectangle: All the properties of the parallelogram;
           4 right angles;
           Diagonals are congruent.
Square: All the properties of the rhombus and the rectangle.

VI. GEOMETRIC MEASUREMENTS

1 yd = 3 ft , 1 ft = 12 in , 1 mile = 5280 ft
1 m = 100 cm , 1 m = 1000 mm

1. Circle

Circumference $C = 2\pi r = \pi d$ r: radius d: diameter
Area $A = \pi r^2$
e.g. When r is doubled, C is doubled and A increases 4 times.

2. Square

Perimeter $P = 4s$ s: length of the side
Area $A = s^2$

3. Rectangle

Perimeter $P = 2l + 2w$ l: length   w: width
Area $A = l \cdot w$

4. Parallelogram

Perimeter P = sum of the 4 sides   b: base   h: height
Area $A = b \cdot h$

5. Trapezoid

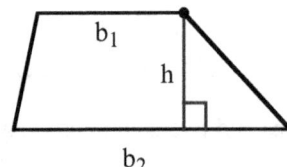

P = sum of 4 sides

$$A = \frac{b_1 + b_2}{2} \cdot h$$

6. Rhombus

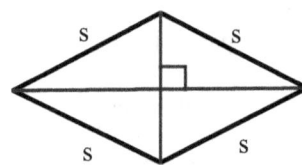

P = 4s

$$A = \frac{1}{2} \cdot d_1 \cdot d_2$$

d_1 and d_2 are diagonals

7. Triangle

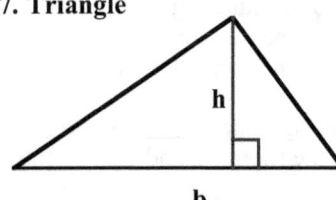

P = sum of 3 sides

$$A = \frac{1}{2} \cdot b \cdot h$$

8. Cube

Volume $V = s^3$ s : side
Surface Area $SA = 6s^2$

9. Rectangular Solid

Volume $V = l \cdot w \cdot h$ l: length, w: width, h: height
Surface Area $SA = 2lw + 2hw + 2\,lh$

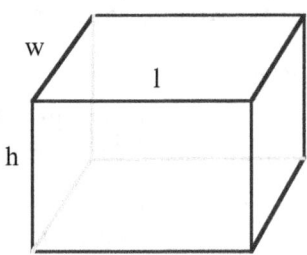

10. Circular Cylinder

Volume $V = B \cdot h$ B: area of the circular base πr^2,
                              h: height
Surface Area $SA = 2\pi r^2 + 2\pi rh$

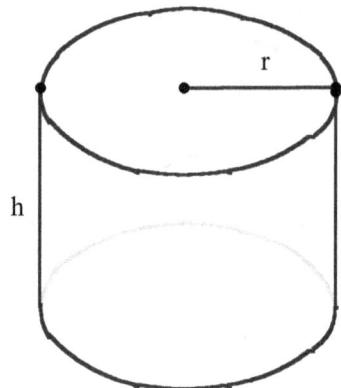

11. Error in Measurement

Absolute Error = |Measured Value - Actual Value|
$$\text{Relative Error} = \frac{\text{Absolute Error}}{\text{Actual Value}}$$
Percent of Error = Relative Error × 100%

e.g. Actual value of the side of a cube is 10.0 cm.
     Measured value is 10.5 cm.
     Find the relative error and percent of error in
     the surface area.

$$\text{Relative Error} = \frac{\left|6(10.5)^2 - 6(10)^2\right|}{6(10)^2} = 0.1025$$

Percent of Error = 0.1025 × 100% = 10.25%

PART 3. Coordinate Geometry and Functions

VII. COORDINATE GEOMETRY AND FUNCTIONS

1. Slope

Coordinate Plane has four Quadrants I, II, III, and IV.

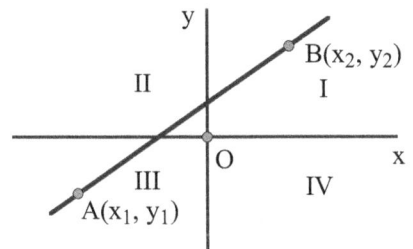

Slope $m = \dfrac{y_2 - y_1}{x_2 - x_1}$

2. Linear Function (First Degree)

A straight line can be represented as a linear function;
The graph of a linear function is a straight line.
Slope-intercept form: $y = mx + b$
where m is the slope and b is the y - intercept.

e.g.

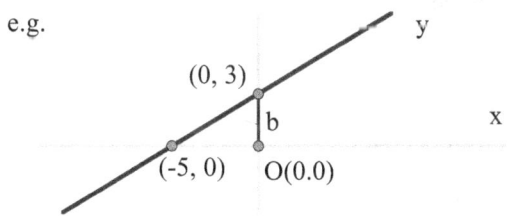

$b = 3, \quad m = \dfrac{3 - 0}{0 - (-5)} = \dfrac{3}{5}$

$y = \dfrac{3}{5} x + 3$

Two parallel lines have the same slope ($m_1 = m_2$);
The slope of a horizontal line is zero (m = 0);
The equation of a horizontal line: $y = b$
The slope of a vertical line is undefined;
The equation of a vertical line: $x = a$

e.g.   Find the slope and y-intercept of $3x - 2y = 12$.
Write the equation in the slope-intercept form:
$y = \dfrac{3}{2} x - 6$,   slope $m = \dfrac{3}{2}$ and y-intercept b = - 6

e.g. Write the equation of a line passing (3, -2) and (6, 4).
First find the slope $m = \dfrac{4 - (-2)}{6 - 3} = \dfrac{6}{3} = 2$
$y = 2x + b$,   replace x by 6 and y by 4
$4 = 2 \cdot 6 + b$ solve for b = - 8
We have the equation of the line $y = 2x - 8$

3. Direct Variation (Ratio and Proportion)

A straight line passing through the Origin is called
Direct Variation:

$$y = mx \qquad or \qquad \dfrac{y}{x} = m$$

The ratio of y to x is called the constant of variation,
which is the slope of the line.

To solve a problem, use $\dfrac{x_1}{x_2} = \dfrac{y_1}{y_2}$

e.g.

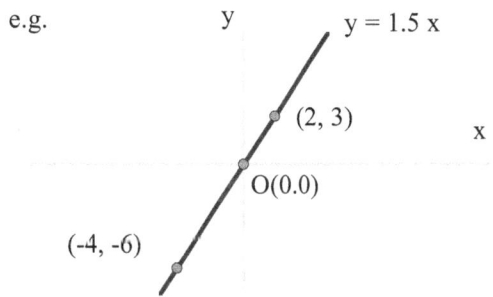

4. Absolute Value Functions

$y = |x|$

when x < 0
y = - x

when x ≥ 0,
y = x

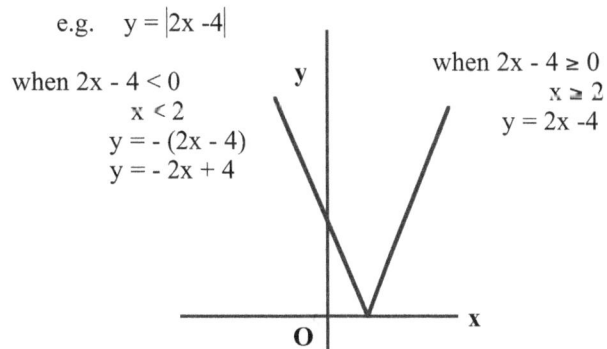

e.g. $y = |2x - 4|$

when 2x - 4 < 0
x < 2
y = - (2x - 4)
y = - 2x + 4

when 2x - 4 ≥ 0
x ≥ 2
y = 2x - 4

5. Quadratic Functions and Parabolas

General Form:

$$y = f(x) = ax^2 + bx + c \qquad \text{where } a \neq 0$$

(1). Axis of Symmetry: $x = -\dfrac{b}{2a}$

(2). Vertex (Turning Point): (x, y)

$$x = -\dfrac{b}{2a}$$

use this value of x to compute $y = ax^2 + bx + c$

(3). Opening:

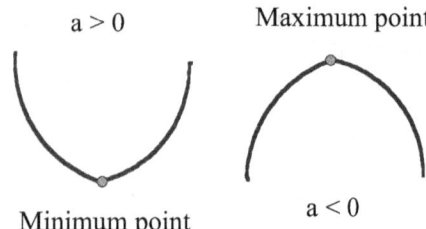

a > 0   Maximum point

Minimum point   a < 0

e.g. $y = x^2 - 2x - 3$

     $a = 1, \ b = -2, \ c = -3$

(1). Axis of Symmetry: $x = \dfrac{-b}{2a} = \dfrac{-(-2)}{2(1)} = 1$

(2). Vertex: $x = 1, \ y = (1)^2 - 2(1) - 3 = -4$

     Vertex: $(1, -4)$

(3). Opening: $a = 1 > 0$, upward

     It has a minimum of -4 at $x = 1$

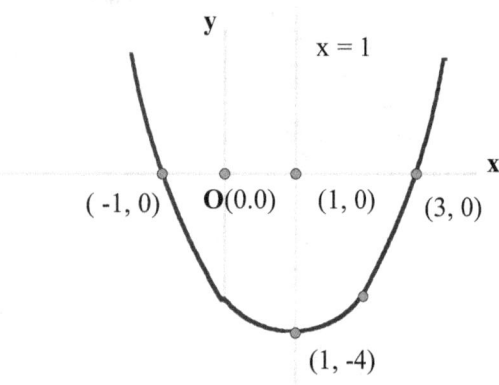

y

x = 1

(-1, 0)   O(0.0)   (1, 0)   (3, 0)   x

(1, -4)

(4). Find the roots of the equation:

     $x^2 - 2x - 3 = 0$

     $(x - 3)(x + 1) = 0$

     $x = -1, \quad x = 3$

   x-intercepts $(-1, 0)$ and $(3, 0)$

6. Exponential Functions

$$y = a^x \qquad \text{where } a > 0 \text{ and } a \neq 1$$

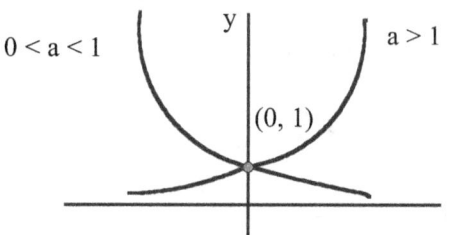

y

$0 < a < 1$ a > 1

(0, 1)

(1). Domain: $\{ x \mid x : \text{all real numbers} \}$

     Range: $\{ y \mid y > 0 \}$

(2). $a > 1$, the function is increasing;

     $a < 1$, the function is decreasing.

(3). when $x = 0, y = 1$

the graphs of the exponential functions passing the point $(0, 1)$

(4). x-axis is the horizontal asymptote.

General Form:

$$y = k \cdot a^x \qquad \text{where } a > 0 \text{ and } a \neq 1$$

$$k \text{ is a constant}$$

Exponential Models:

e.g. $A = A_0 2^{\frac{t}{96}}$

     A_0 is the original amount. (when $t = 0$)

     A is the amount at time t.

If the original amount $A_0 = 250$,

find the amount A when the time is 24.

$$A = 250 \cdot 2^{\frac{24}{96}}$$

$$A = 250 \cdot 2^{0.25} = 297.3$$

e.g. A new car will depreciate at a rate of 8% per year. If a new car is worth $15,000, how much will it be worth after 3 years?

$$A = A_0(1 - 8\%)^t$$

$$= 15{,}000(0.92)^3$$

$$= 11{,}680$$

Answer: The car will be worth $11,680 after 3 years.

7. Graphic Solutions of System of Equations

Linear System:

e.g. $x + y = 7$
$2x - y = 2$
rewrite in the slope-intercept form
$y = -x + 7$
$y = 2x - 2$

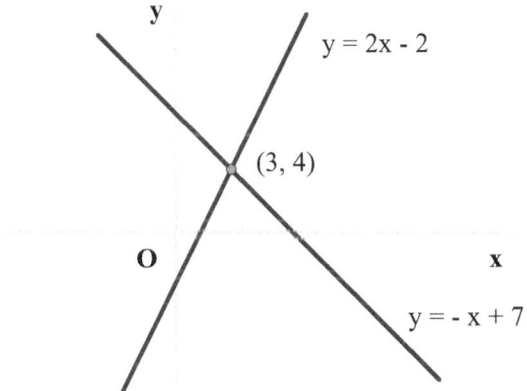

The solution of the system of equations is the intersection point (3, 4) of the two lines.

Quadratic-Linear System

e.g. $y = x^2 - 8$ (1)
$y + 5 = 2x$ (2)
rewrite Eq. (2) in the slope-intercept form
$y = 2x - 5$

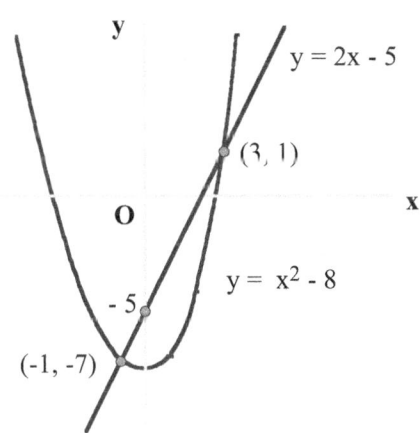

The solution of the system of equations are the intersection points (3, 1) and (-1, -7)

8. Linear Inequalities

e.g. The solution of $y < x + 2$ is the region under $y = x + 2$;

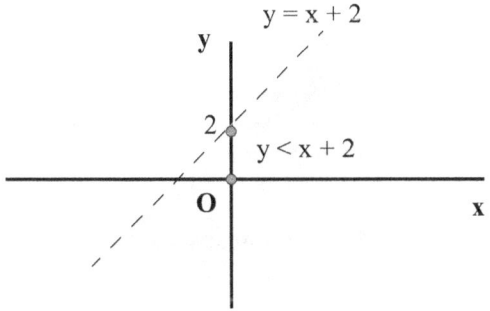

e.g. The solution of $y \geq x + 2$ is the region above $y = x + 2$ and including the line $y = x + 2$

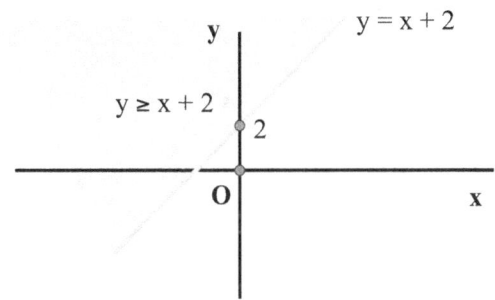

The System of Linear Inequalities

e.g. The solution of the system
$x > 2$ (1)
$2x - y \geq 6$ (2)
rewrite (2) in the slope-intercept form
$-y \geq -2x + 6$
$y \leq 2x - 6$ inequality sign reversed

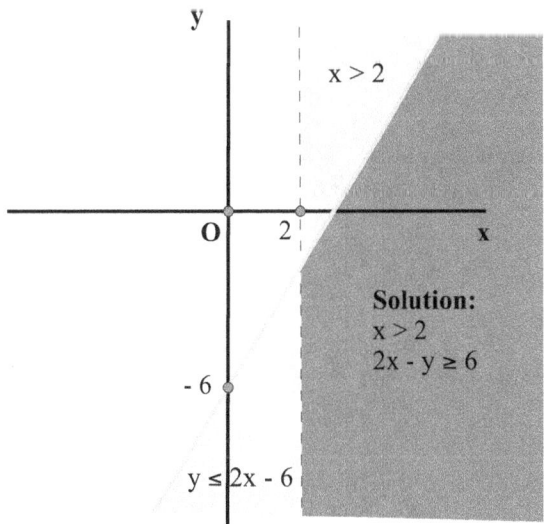

9. Relations and Functions

Relation (Not a Function)   Function

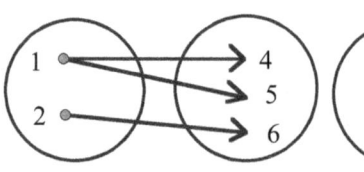
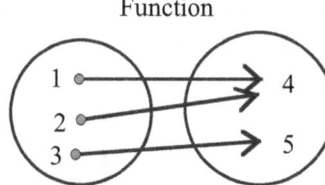

A function is a special relation.
The first element in the ordered pairs can not repeat in a function.
e.g. $\{(2, 1), (3, 1), (4, 3), (5, 4)\}$ is a function.
   $\{(1, 2), (1, 3), (3, 4), (4, 5)\}$ is not a function.

The **Function Notation** of $y = x^2 + 1$ is $f(x) = x^2 + 1$
   $f(2) = (2)^2 + 1 = 5$

Domain and Range:

e.g. $y = x^2$ Domain: $\{x \mid x$ all real numbers$\}$
                      Range: $\{y \mid y \geq 0\}$
e.g. $y = \sqrt{x}$ Domain: $\{x \mid x \geq 0\}$
                      Range: $\{y \mid y \geq 0\}$
e.g. $y = \dfrac{1}{x^2 - 9}$ Domain: $\{x \mid x$ all real numbers except $\pm 3\}$

e.g. $y = \dfrac{1}{\sqrt{x - 3}}$ Domain: $\{x \mid x > 3\}$

Vertical Line Test:

If any vertical line intersects the graph at only one point, then the relation is a function.
e.g.   Graph $y = x^2$ is a function

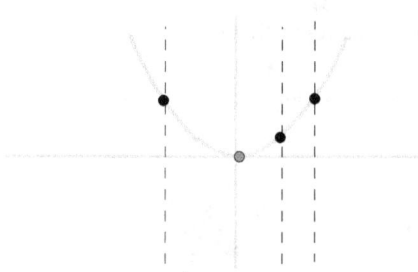

$y^2 = x$ is equivalent to $y = \pm\sqrt{x}$. It is not a function.

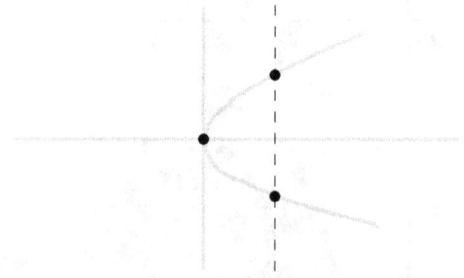

PART 4. Probability and Statistics

VIII. PROBABILITY

1. Probability and Counting Principle

Sample Space: all possible outcomes
Event: the favorable outcomes

(1) The probability of a simple event
$$P(E) = \frac{\text{number of the outcomes of the event}}{\text{nubmer of the outcomes of the sample space}}$$
$$P(E) = \frac{n(E)}{n(S)}$$
e.g. A bag contains 6 black balls and 4 white balls.
   What is the probability of selecting a black ball?
$$P(\text{Black}) = \frac{n(\text{Black})}{n(\text{Sample Space})} = \frac{6}{10}$$

Impossible Case: $P(E) = 0$
Certain Case: $P(E) = 1$
Negation: $P(\text{Not E}) = 1 - P(E)$

e.g. If $P(\text{rain}) = 30\%$, then $P(\text{Not rain}) = 1 - P(\text{rain}) = 70\%$

(2) The probability of a single event with two conditions:
   $P(A \text{ and } B)$ meets both conditions
e.g. 52 cards , $P(K \text{ and red}) = \dfrac{2}{52}$

(3) The probability of a single event that satisfies
   condition A or condition B
      $P(A \text{ or } B) = P(A) + P(B) - P(A \text{ and } B)$
e.g. 52 cards , $P(K \text{ or red})$
      $= P(K) + P(\text{red}) - P(K \text{ and red})$
      $= \dfrac{4}{52} + \dfrac{26}{52} - \dfrac{2}{52}$
      $= \dfrac{28}{52}$
For disjoint sets A and B,
we have $P(A \text{ and } B) = 0$, then $P(A \text{ or } B) = P(A) + P(B)$
e.g.   52 cards , $P(K \text{ or } J)$
      $= P(K) + P(J)$
      $= \dfrac{4}{52} + \dfrac{4}{52}$
      $= \dfrac{8}{52}$
      here $P(K \text{ and } J) = 0$

(4) **Counting Principle** (2 or more activities)

If the first activity can occur in M ways and the second activity can occur in N ways, then both activities can occur in M•N ways.

e.g. 3 doors to a building, 2 stairways to the second floor. There are 3•2 = 6 different ways to go.

Tree Diagram:

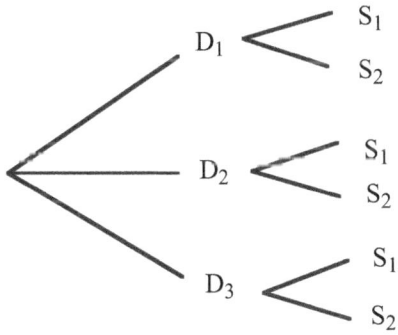

Sample Space:
$\{(D_1, S_1), (D_1, S_2), (D_2, S_1), (D_2, S_2), (D_3, S_1), (D_3, S_2)\}$

(5) Counting Principle for Probability

When A and B are indepenent events, the compound event of A and B has the probability
$$P(A, B) = P(A) \cdot P(B)$$

e.g. 4 students. The probability of the tallest one in the first place (A) and the shortest one in the last place (B)
$$P(A, B) = P(A) \cdot P(B) = \frac{1}{4} \cdot \frac{1}{3} = \frac{1}{12}$$

2. Permutation and Combination

In a **permutation** the order of the objects is important.
(1) The permutaion of n objects taken n at a time
$$_nP_n = n! = n(n-1)(n-2) \ldots 2 \cdot 1$$

e.g. Five letters A, B, C, D, E have 5! different arrangements. (5! = 5•4•3•2•1 = 120)

(2) The permutation of n objects taken n at a time with r items identical
$$\frac{n!}{r!}$$

e.g. Five letters COLOR have $\frac{5!}{2!}$ different arrangements.
$$(\frac{5!}{2!} = \frac{120}{2 \cdot 1} = 60)$$

(3) The permutation of n objects taken r (r < n) at a time
$$_nP_r = n(n-1)(n-2)\ldots \qquad (\text{r factors})$$

e.g. How many different arrangements of 1st, 2nd, and 3rd place are possible for 10 students?
$$_{10}P_3 = 10 \cdot 9 \cdot 8 = 720 \qquad (3 \text{ factors})$$

In a **combination** the order of the objects does not matter.

e.g. (A, B, C) and (C, B, A) are considered same.

(4) The combination of n objects taken r at a time
$$_nC_r = \frac{_nP_r}{r!} \qquad (r \leq n)$$
$$_nC_n = 1, \quad _nC_0 = 1, \quad _nC_1 = n, \quad _nC_r = {_nC_{n-r}}$$

e.g. How many 3 player teams can be formed from 10 students?
$$_{10}C_3 = \frac{_{10}P_3}{3!} = \frac{10 \cdot 9 \cdot 8}{3 \cdot 2 \cdot 1} = 120$$

e.g. $_{50}C_{48} = {_{50}C_2}$ (to simplify the calculation)

IX. STATISTICS

Quantitative Data is measurable or countable.
Qualitative Data is unmeasurable or uncountable.

Collect Data:
The sample must be large enough to be effective and must be chosen **randomly** to eliminate any **bias**.

1. Statistics (Univariate Data)

Analyze Data:
First arrange the data in numerical order.

$$\text{Mean} = \text{Average} = \frac{\text{sum of the data values}}{\text{number of the data items}}$$

Median: the middle value when the data arranged in order
Mode: the value that appears most often
Range: the difference between the highest value and the lowest value
Percentile: a number that tells what percent of the total number of the data values are less than or equal to a given data point
1st Quartile (25th percentile): the middle value of the lower half set of the data, aka. **Lower Quartile**
2nd Quartile (50th percentile): the median, aka. **Middle Quartile**
3rd Quartile (75th percentile): the middle value of the upper half set of the data, aka. **Upper Quartile**

Outliers: Some data points far outside most of the points in the data set.
Outliers can strongly affect the mean value. When outliers exist, use median to represent the central tendency of the data.

e.g. Analyze the grades:
    78, 85, 81, 95, 61, 85, 75, 88, 72, 100
First rearrange the data in numerical order:
    61, 72, 75, 78, 81, 85, 85, 88, 95, 100
(make sure the number of items are same)

$$\text{Mean} = \frac{820}{10} = 82$$

$$\text{Median} = \frac{81 + 85}{2} = 83$$

(if the set has an even number of data values, take the average of the two middle values)
Mode = 85
Range = 100 - 61 = 39
Middle Quartile = Median = 83
Lower Quartile = 75
Upper Quartile = 88

Box-and-Whisker Plot :

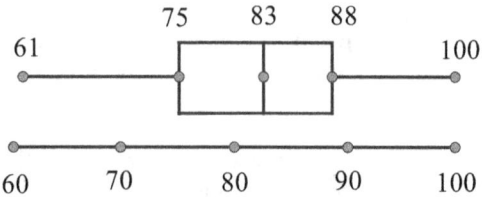

Frequency Table :

Interval	Frequency
61 - 70	1
71 - 80	3
81- 90	4
91 - 100	2

Frequency Histogram :

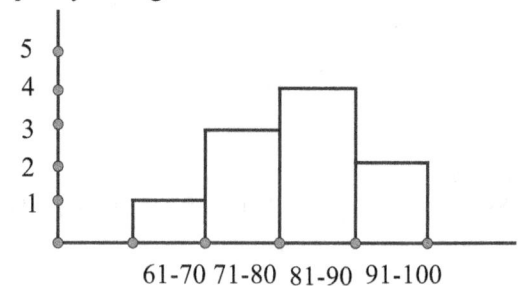

Cumulative Frequency Table :

Interval	Cumulative Frequency
61 - 70	1
61 - 80	4
61- 90	8
61 - 100	10

Cumulative Frequency Histogram :

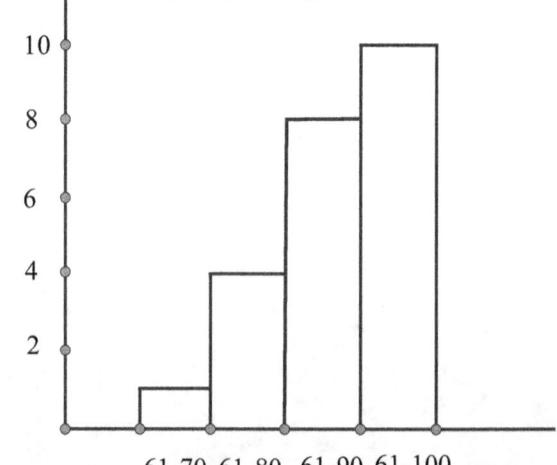

2. Statistics (Bivariate Data)

Correlation: The relationship between two sets
      of data
Causation: The relationship in which one variable
      produces an effect on the other

Regression Modeling
Linear Regression: $y = ax + b$

Correlation Coefficient r :

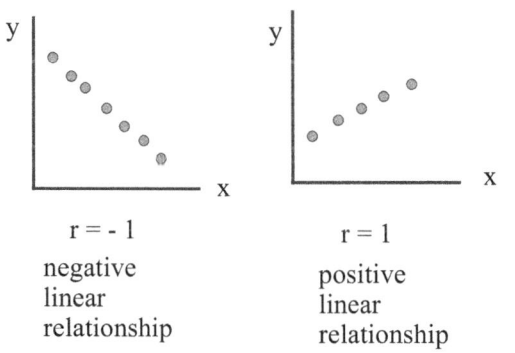

$r = -1$

negative
linear
relationship

$r = 1$

positive
linear
relationship

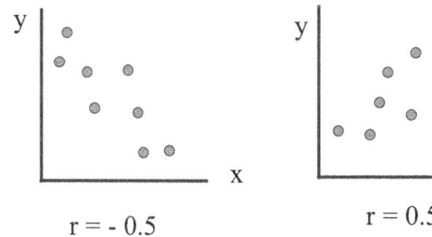

$r = -0.5$

moderate
negative linear
relationship

$r = 0.5$

moderate
positive linear
relationship

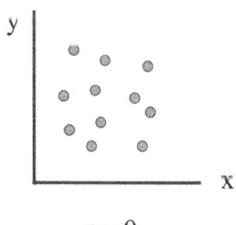

$r = 0$

no linear
relationship

Line of Best Fit (The Linear Regression)

(1). passing through the mean point $(\overline{x}, \overline{y})$
(2). the difference between the model values and the
    real values is the least

Use graphing calculator to find the equation of the
Line of Best Fit : $y = ax + b$

e.g.

x_i	2	4	6	8	10
y_i	13	15	16	17	20

(1) Clear List L_1 and List L_2
[STAT] EDIT / 4: ClrList [ENTER] [2nd] [L_1] [,]
[2nd] [L_2] [ENTER]
(2) Enter data to L_1 and L_2
[STAT] EDIT / 1: Edit ... [ENTER]
Enter data x_i into List L_1 ;   Enter data y_i into List L_2 .
(3) Scatter Plot: [2nd] [STAT PLOT] 1: PLOT 1 [ENTER]

    ON
    Type:

[ZOOM] [9]

(4) Find the equation of the Line of Best Fit
    and the Correlation Coefficient r :
[2nd] [CATALOG] Diagnostic On [ENTER]
[STAT] CALC / 4: LinReg(ax + b) [ENTER]
[2nd] [L_1] [,] [L_2] [ENTER]
LinReg $y = ax + b$
    $a = 0.8$ $b = 11.4$ $r = 0.98$

(5) Draw the Line of Best Fit:
[Y =] [VARS] 5: Statistics ... [ENTER] EQ /
1: RegEQ [ENTER] [ZOOM] [9]

(6) Predict the results by using the model:
Find the value of y when x = 10.5
[2nd] [CALC] 1: Values [ENTER]
X = 10.5 [ENTER]   Y = 19.8
(Adjust Window Dimensions for Extrapolation)

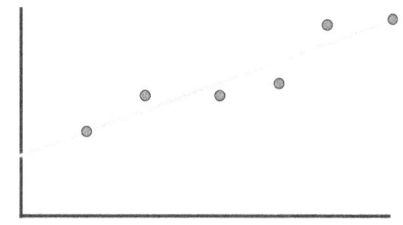

Reference Sheet

Trigonometric Ratios	$\sin A = \dfrac{opposite}{hypotenuse}$
	$\cos A = \dfrac{adjacent}{hypotenuse}$
	$\tan A = \dfrac{opposite}{adjacent}$

Area	trapezoid	$A = \frac{1}{2}h(b_1 + b_2)$

Volume	cylinder	$V = \pi r^2 h$

Surface Area	rectangular prism	$SA = 2lw + 2hw + 2lh$
	cylinder	$SA = 2\pi r^2 + 2\pi rh$

Coordinate Geometry	$m = \dfrac{\boldsymbol{\Delta} y}{\boldsymbol{\Delta} x} = \dfrac{y_2 - y_1}{x_2 - x_1}$

1. Which interval notation represents the set of all numbers from 2 through 7, inclusive?
(1)   (2, 7]   (2)   (2, 7)   (3)   [2, 7)   (4)   [2, 7]

2. The set $\{1, 2, 3, 4\}$ is equivalent to
(1) $\left\{x \mid 1 < x < 4, \text{ where } x \text{ is a whole number}\right\}$ (3) $\left\{x \mid 0 < x \leq 4, \text{ where } x \text{ is a whole number}\right\}$
(2) $\left\{x \mid 0 < x < 4, \text{ where } x \text{ is a whole number}\right\}$ (4) $\left\{x \mid 1 < x \leq 4, \text{ where } x \text{ is a whole number}\right\}$

3. The set $\{11, 12\}$ is equivalent to
(1) $\{x \mid 11 < x < 12, \text{ where } x \text{ is an integer}\}$ (3) $\{x \mid 10 \leq x < 12, \text{ where } x \text{ is an integer}\}$
(2) $\{x \mid 11 < x \leq 12, \text{ where } x \text{ is an integer}\}$ (4) $\{x \mid 10 < x \leq 12, \text{ where } x \text{ is an integer}\}$

4. Which interval notation represents the set of all numbers greater than or equal to 5 and less than 12?
(1)   [5, 12)   (2)   (5, 12]   (3)   (5, 12)   (4)   [5, 12]

5. Given: Set $A = \{(-2, -1), (-1, 0), (1, 8)\}$

   Set $B = \{(-3, -4), (-2, -1), (-1, 2), (1, 8)\}$.
   What is the intersection of sets A and B?
(1) $\{(1, 8)\}$ (3) $\{(-2, -1), (1, 8)\}$
(2) $\{(-2, -1)\}$ (4) $\{(-3, -4), (-2, -1), (-1, 2), (-1, 0), (1, 8)\}$

6. Consider the set of integers greater than -2 and less than 6. A subset of this set is the positive factors of 5. What is the complement of this subset?
(1) $\{0, 2, 3, 4\}$ (3) $\{-2, -1, 0, 2, 3, 4, 6\}$
(2) $\{-1, 0, 2, 3, 4\}$ (4) $\{-2, -1, 0, 1, 2, 3, 4, 5, 6\}$

7. Given: $Q = \{0, 2, 4, 6\}$

   $W = \{0, 1, 2, 3\}$

   $Z = \{1, 2, 3, 4\}$
What is the intersection of sets Q, W, and Z?
(1) $\{2\}$ (2) $\{0, 2\}$ (3) $\{1, 2, 3\}$ (4) $\{0, 1, 2, 3, 4, 6\}$

8. Given: Set $U = \{S, O, P, H, I, A\}$

   Set $B = \{A, I, O\}$
If set B is a subset of set U, what is the complement of set B?
(1) $\{O, P, S\}$ (2) $\{I, P, S\}$ (3) $\{A, H, P\}$ (4) $\{H, P, S\}$

9. The value of the expression $-|a - b|$ when $a = 7$ and $b = -3$ is
(1)   -10   (2)   10   (3)   -4   (4)   4

10. Which property is illustrated by the equation $ax + ay = a(x + y)$?
(1)   associative   (2)   commutative   (3)   distributive   (4)   identity

I. Sets, Numbers, and Operations

11. What is the additive inverse of the expression $a - b$?

(1) $a + b$ (2) $a - b$ (3) $-a + b$ (4) $-a - b$

12. Debbie solved the linear equation $3(x + 4) - 2 = 16$ as follows:

[Line 1] $3(x + 4) - 2 = 16$
[Line 2] $3(x + 4) = 18$
[Line 3] $3x + 4 = 18$
[Line 4] $3x = 14$
[Line 5] $x = 4\frac{2}{3}$

She made an error between lines

(1)   1 and 2   (2)   2 and 3   (3)   3 and 4   (4)   4 and 5

Show Work:

1. Maureen tracks the range of outdoor temperatures over three days. She records the following information.

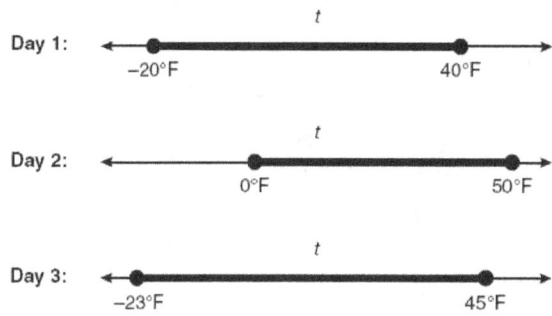

Express the intersection of the three sets as an inequality in terms of temperature, t.

2. Perform the indicated operation: $-6(a - 7)$
State the name of the property used.

3. The Hudson Record Store is having a going-out-of-business sale. CDs normally sell for $18.00. During the first week of the sale, all CDs will sell for $15.00. Written as a fraction, what is the rate of discount? What is this rate expressed as a percent? Round your answer to the *nearest hundredth of a percent*. During the second week of the sale, the same CDs will be on sale for 25% off the *original* price. What is the price of a CD during the second week of the sale?

1. What is the product of $-3x^2y$ and $(5xy^2 + xy)$?
(1) $-15x^3y^3 - 3x^3y^2$ (3) $-15x^2y^2 - 3x^2y$
(2) $-15x^3y^3 - 3x^3y$ (4) $-15x^3y^3 + xy$

2. When $4x^2 + 7x - 5$ is subtracted from $9x^2 - 2x + 3$, the result is
(1) $5x^2 + 5x - 2$ (3) $-5x^2 + 5x - 2$
(2) $5x^2 - 9x + 8$ (4) $-5x^2 + 9x - 8$

3. The sum of $4x^3 + 6x^2 + 2x - 3$ and $3x^3 + 3x^2 - 5x - 5$ is
(1) $7x^3 + 3x^2 - 3x - 8$ (3) $7x^3 + 9x^2 - 3x - 8$
(2) $7x^3 + 3x^2 + 7x + 2$ (4) $7x^6 + 9x^4 - 3x^2 - 8$

4. Factored, the expression $16x^2 - 25y^2$ is equivalent to
(1) $(4x - 5y)(4x + 5y)$ (3) $(8x - 5y)(8x + 5y)$
(2) $(4x - 5y)(4x - 5y)$ (4) $(8x - 5y)(8x - 5y)$

5. If Ann correctly factors an expression that is the difference of two perfect squares, her factors could be
(1) $(2x + y)(x - 2y)$ (3) $(x - 4)(x - 4)$
(2) $(2x + 3y)(2x - 3y)$ (4) $(2y - 5)(y - 5)$

6. Factored completely, the expression $3x^2 - 3x - 18$ is equivalent to
(1) $3(x^2 - x - 6)$ (3) $(3x - 9)(x + 2)$
(2) $3(x - 3)(x + 2)$ (4) $(3x + 6)(x - 3)$

7. Which expression represents $\dfrac{(2x^3)(8x^5)}{4x^6}$ in simplest form?

(1) x^2 (2) x^9 (3) $4x^2$ (4) $4x^9$

8. The expression $\dfrac{9x^4 - 27x^6}{3x^3}$ is equivalent to

(1) $3x(1 - 3x)$ (3) $3x(1 - 9x^5)$
(2) $3x(1 - 3x^2)$ (4) $9x^3(1 - x)$

9. What is the product of $\dfrac{x^2 - 1}{x + 1}$ and $\dfrac{x + 3}{3x - 3}$ expressed in simplest form?

(1) x (2) $\dfrac{x}{3}$ (3) $x + 3$ (4) $\dfrac{x + 3}{3}$

10. Which expression represents $\dfrac{2x^2 - 12x}{x - 6}$ in simplest form?

(1) 0 (2) $2x$ (3) $4x$ (4) $2x + 2$

II. Algebraic Expressions and Operations

11. What is $\dfrac{6}{5x} - \dfrac{2}{3x}$ in simplest form?

(1) $\dfrac{8}{15x^2}$　　　(2) $\dfrac{8}{15x}$　　　(3) $\dfrac{4}{15x}$　　　(4) $\dfrac{4}{2x}$

12. The function $y = \dfrac{x}{x^2 - 9}$ is undefined when the value of x is

(1) 0 or 3　　　(2) 3 or -3　　　(3) 3, only　　　(4) -3, only

13. Which value of n makes the expression $\dfrac{5n}{2n - 1}$ undefined?

(1) 1　　　(2) 0　　　(3) $-\dfrac{1}{2}$　　　(4) $\dfrac{1}{2}$

14. Which expression represents $\dfrac{x^2 - 2x - 15}{x^2 + 3x}$ in simplest form?

(1) -5　　　(2) $\dfrac{x - 5}{x}$　　　(3) $\dfrac{-2x - 5}{x}$　　　(4) $\dfrac{-2x - 15}{3x}$

15. What is $\dfrac{6}{4a} - \dfrac{2}{3a}$ expressed in simplest form?

(1) $\dfrac{4}{a}$　　　(2) $\dfrac{5}{6a}$　　　(3) $\dfrac{8}{7a}$　　　(4) $\dfrac{10}{12a}$

16. Which expression represents $\dfrac{-14a^2 c^8}{7a^3 c^2}$ in simplest form?

(1) $-2ac^4$　　　(2) $-2ac^6$　　　(3) $\dfrac{-2c^4}{a}$　　　(4) $\dfrac{-2c^6}{a}$

17. What is the sum of $\dfrac{-x + 7}{2x + 4}$ and $\dfrac{2x + 5}{2x + 4}$?

(1) $\dfrac{x + 12}{2x + 4}$　　　(2) $\dfrac{3x + 12}{2x + 4}$　　　(3) $\dfrac{x + 12}{4x + 8}$　　　(4) $\dfrac{3x + 12}{4x + 8}$

18. What is $\dfrac{\sqrt{32}}{4}$ expressed in simplest radical form?

(1) $\sqrt{2}$　　　(2) $4\sqrt{2}$　　　(3) $\sqrt{8}$　　　(4) $\dfrac{\sqrt{8}}{2}$

19. The expression $6\sqrt{50} + 6\sqrt{2}$ written in simplest radical form is

(1) $6\sqrt{52}$　　　(2) $12\sqrt{52}$　　　(3) $17\sqrt{2}$　　　(4) $36\sqrt{2}$

20. The expression $\sqrt{72} - 3\sqrt{2}$ written in simplest radical form is

(1) $5\sqrt{2}$ (2) $3\sqrt{6}$ (3) $3\sqrt{2}$ (4) $\sqrt{6}$

21. What is half of 2^6?

(1) 1^3 (2) 1^6 (3) 2^3 (4) 2^5

22. Which expression is equivalent to $(3x^2)^3$?

(1) $9x^5$ (2) $9x^6$ (3) $27x^5$ (4) $27x^6$

23. Which expression is equivalent to $3^3 \bullet 3^4$?

(1) 9^{12} (2) 9^7 (3) 3^{12} (4) 3^7

24. What is the product of 8.4×10^8 and 4.2×10^3 written in scientific notation?

(1) 2.0×10^5 (2) 12.6×10^{11} (3) 35.28×10^{11} (4) 3.528×10^{12}

25. What is the product of 12 and 4.2×10^6 expressed in scientific notation?

(1) 50.4×10^6 (2) 50.4×10^7 (3) 5.04×10^6 (4) 5.04×10^7

26. The length of a rectangular room is 7 less than three times the width, w, of the room. Which expression represents the area of the room?

(1) $3w - 4$ (2) $3w - 7$ (3) $3w^2 - 4w$ (4) $3w^2 - 7w$

27. If $a + ar = b + r$, the value of a in terms of b and r can be expressed as

(1) $\dfrac{b}{r} + 1$ (2) $\dfrac{1 + b}{r}$ (3) $\dfrac{b + r}{1 + r}$ (4) $\dfrac{1 + b}{r + b}$

28. An example of an algebraic expression is

(1) $\dfrac{2x + 3}{7} = \dfrac{13}{x}$ (3) $4x - 1 = 4$

(2) $(2x + 1)(x - 7)$ (4) $x = 2$

29. Which verbal expression is represented by $\dfrac{1}{2}(n - 3)$?

(1) one-half n decreased by 3   (3) the difference of one-half n and 3
(2) one-half n subtracted from 3   (4) one-half the difference of n and 3

30. A formula used for calculating velocity is $v = \dfrac{1}{2}at^2$. What is a expressed in terms of v and t?

(1) $a = \dfrac{2v}{t}$ (2) $a = \dfrac{2v}{t^2}$ (3) $a = \dfrac{v}{t}$ (4) $a = \dfrac{v}{2t^2}$

22. **II. Algebraic Expressions and Operations**

Show Work:

1. Factor completely: $4x^3 - 36x$

2. Simplify: $\dfrac{27k^5 m^8}{(4k^3)(9m^2)}$

3. Perform the indicated operation and simplify: $\dfrac{3x+6}{4x+12} \div \dfrac{x^2-4}{x+3}$

4. Express in simplest form: $\dfrac{x^2+9x+14}{x^2-49} \div \dfrac{3x+6}{x^2+x-56}$

5. Express $5\sqrt{72}$ in simplest radical form.

1. Pam is playing with red and black marbles. The number of red marbles she has is three more than twice the number of black marbles she has. She has 42 marbles in all. How many red marbles does Pam have?
(1)   13   (2)   15   (3)   29   (4)   33

2. Sam and Odel have been selling frozen pizzas for a class fundraiser. Sam has sold half as many pizzas as Odel. Together they have sold a total of 126 pizzas. How many pizzas did Sam sell?
(1)   21   (2)   42   (3)   63   (4)   84

3. What is the speed, in meters per second, of a paper airplane that flies 24 meters in 6 seconds?
(1)   144   (2)   30   (3)   18   (4)   4

4. Tamara has a cell phone plan that charges $0.07 per minute plus a monthly fee of $19.00. She budgets $29.50 per month for total cell phone expenses without taxes. What is the maximum number of minutes Tamara could use her phone each month in order to stay within her budget?
(1)   150   (2)   271   (3)   421   (4)   692

5. Rhonda has $1.35 in nickels and dimes in her pocket. If she has six more dimes than nickels, which equation can be used to determine x, the number of nickels she has?
(1) $0.05(x+6)+0.10x = 1.35$ (3) $0.05+0.10(6x) = 1.35$
(2) $0.05x+0.10(x+6) = 1.35$ (4) $0.15(x+6) = 1.35$

6. If the speed of sound is 344 meters per second, what is the approximate speed of sound, in meters per hour?

> 60 seconds = 1 minute
> 60 minutes = 1 hour

(1)   20,640   (2)   41,280   (3)   123,840   (4)   1,238,400

7. At Genesee High School, the sophomore class has 60 more students than the freshman class. The junior class has 50 fewer students than twice the students in the freshman class. The senior class is three times as large as the freshman class. If there are a total of 1,424 students at Genesee High School, how many students are in the freshman class?
(1)   202   (2)   205   (3)   235   (4)   236

8. Steve ran a distance of 150 meters in $1\frac{1}{2}$ minutes. What is his speed in meters per hour?
(1)   6   (2)   60   (3)   100   (4)   6,000

9. An electronics store sells DVD players and cordless telephones. The store makes a $75 profit on the sale of each DVD player (d) and a $30 profit on the sale of each cordless telephone (c). The store wants to make a profit of at least $255.00 from its sales of DVD players and cordless phones. Which inequality describes this situation?
(1) $75d+30c < 255$ (3) $75d+30c > 255$
(2) $75d+30c \le 255$ (4) $75d+30c \ge 255$

10. Which value of x is in the solution set of the inequality $-2x+5 > 17$?
(1)   −8   (2)   −6   (3)   −4   (4)   12

11. Students in a ninth grade class measured their heights, h, in centimeters. The height of the shortest student was 155 cm, and the height of the tallest student was 190 cm. Which inequality represents the range of heights?

(1) $155 < h < 190$ (3) $h \geq 155$ or $h \leq 190$

(2) $155 \leq h \leq 190$ (4) $h > 155$ or $h < 190$

12. The sign shown below is posted in front of a roller coaster ride at the Wadsworth County Fairgrounds.

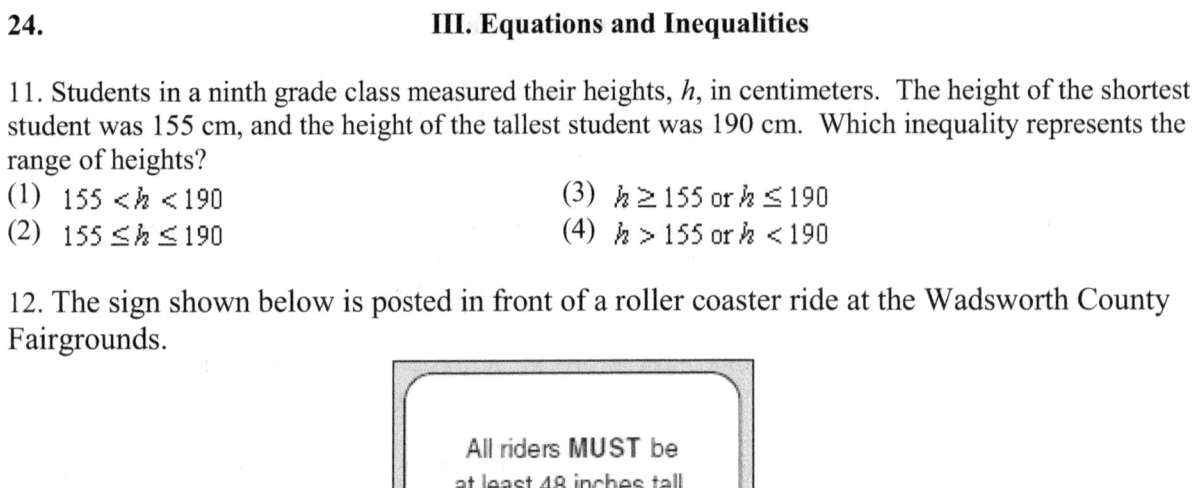

All riders **MUST** be
at least 48 inches tall.

If h represents the height of a rider in inches, what is a correct translation of the statement on this sign?

(1) $h < 48$ (2) $h > 48$ (3) $h \leq 48$ (4) $h \geq 48$

13. Which value of x is in the solution set of $\frac{4}{3}x + 5 < 17$?

(1)   8   (2)   9   (3)   12   (4)   16

14. The length of a rectangular window is 5 feet more than its width, w. The area of the window is 36 square feet. Which equation could be used to find the dimensions of the window?

(1) $w^2 + 5w + 36 = 0$ (3) $w^2 - 5w + 36 = 0$

(2) $w^2 - 5w - 36 = 0$ (4) $w^2 + 5w - 36 = 0$

15. What are the roots of the equation $x^2 - 10x + 21 = 0$?

(1) 1 and 21   (3) 3 and 7

(2) -5 and -5 (4) -3 and -7

16. When 36 is subtracted from the square of a number, the result is five times the number. What is the positive solution?

(1)   9   (2)   6   (3)   3   (4)   4

17. Which value of x is a solution of $\frac{5}{x} = \frac{x+13}{6}$?

(1)   -2   (2)   -3   (3)   -10   (4)   -15

18. What is the value of x in the equation $\frac{2}{x} - 3 = \frac{26}{x}$?

(1) -8 (2) $-\frac{1}{8}$ (3) $\frac{1}{8}$ (4)   8

19. Which value of x is the solution of $\frac{x}{3} + \frac{x+1}{2} = x$?
(1)   1   (2)   -1   (3)   3   (4)   -3

20. The sum of two numbers is 47, and their difference is 15. What is the larger number?
(1)   16   (2)   31   (3)   32   (4)   36

21. Jack bought 3 slices of cheese pizza and 4 slices of mushroom pizza for a total cost of $12.50. Grace bought 3 slices of cheese pizza and 2 slices of mushroom pizza for a total cost of $8.50. What is the cost of one slice of mushroom pizza?
(1) $1.50   (2) $2.00   (3) $3.00   (4) $3.50

22. What is the value of the y-coordinate of the solution to the system of equations $x + 2y = 9$ and $x - y = 3$?
(1)   6   (2)   2   (3)   3   (4)   5

23. Julia went to the movies and bought one jumbo popcorn and two chocolate chip cookies for $5.00. Marvin went to the same movie and bought one jumbo popcorn and four chocolate chip cookies for $6.00. How much does one chocolate chip cookie cost?
(1) $0.50   (2) $0.75   (3) $1.00   (4) $2.00

24. Which ordered pair is a solution of the system of equations $y = x^2 - x - 20$ and $y = 3x - 15$?
(1) $(-5, -30)$ (2) $(-1, -18)$ (3) $(0, 5)$ (4) $(5, -1)$

26.

III. Equations and Inequalities

Show Work:

1. Hannah took a trip to visit her cousin. She drove 120 miles to reach her cousin's house and the same distance back home. It took her 1.2 hours to get halfway to her cousin's house. What was her average speed, in miles per hour, for the first 1.2 hours of the trip? Hannah's average speed for the remainder of the trip to her cousin's house was 40 miles per hour. How long, in hours, did it take her to drive the remaining distance? Traveling home along the same route, Hannah drove at an average rate of 55 miles per hour. After 2 hours her car broke down. How many miles was she from home?

2. Peter begins his kindergarten year able to spell 10 words. He is going to learn to spell 2 new words every day. Write an inequality that can be used to determine how many days, d, it takes Peter to be able to spell *at least* 75 words. Use this inequality to determine the minimum number of whole days it will take for him to be able to spell *at least* 75 words.

3. Given: $A = \{18, 6, -3, -12\}$

Determine all elements of set A that are in the solution of the inequality $\frac{2}{3}x + 3 < -2x - 7$.

4. A contractor needs 54 square feet of brick to construct a rectangular walkway. The length of the walkway is 15 feet more than the width. Write an equation that could be used to determine the dimensions of the walkway. Solve this equation to find the length and width, in feet, of the walkway.

5. Find three consecutive positive even integers such that the product of the second and third integers is twenty more than ten times the first integer. [Only an algebraic solution can receive full credit.]

6. Solve for x: $\dfrac{x+1}{x} = \dfrac{-7}{x-12}$

7. Solve the following system of equations algebraically:
$$3x + 2y = 4$$
$$4x + 3y = 7$$
[Only an algebraic solution can receive full credit.]

28.   **IV. Right Triangles and Trigonometry**

1. What is the value of *x*, in inches, in the right triangle below?

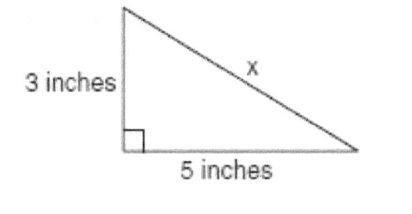

(1) $\sqrt{15}$ (2) 8   (3) $\sqrt{34}$ (4) 4

2. The center pole of a tent is 8 feet long, and a side of the tent is 12 feet long as shown in the diagram below.

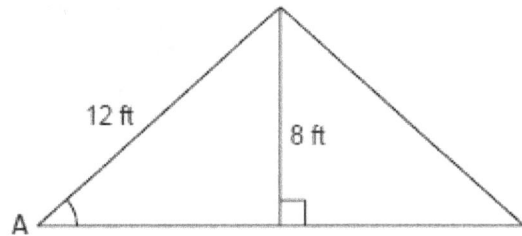

If a right angle is formed where the center pole meets the ground, what is the measure of angle *A* to the *nearest degree*?

(1) 34   (2) 42   (3) 48   (4) 56

3. In the right triangle shown in the diagram below, what is the value of *x* to the *nearest whole number*?

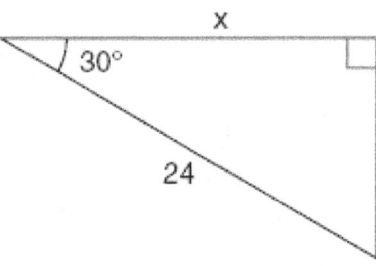

(1) 12   (2) 14   (3) 21   (4) 28

4. The diagram below shows right triangle *UPC*.

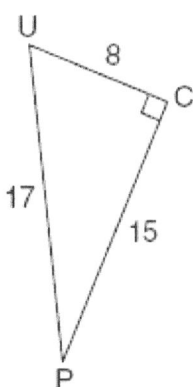

Which ratio represents the sine of ∠*U*?

(1) $\dfrac{15}{8}$ (2) $\dfrac{15}{17}$ (3) $\dfrac{8}{15}$ (4) $\dfrac{8}{17}$

5. Which equation shows a correct trigonometric ratio for angle *A* in the right triangle below?

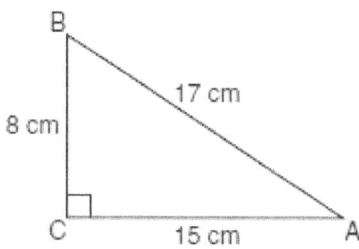

(1) $\sin A = \dfrac{15}{17}$ (2) $\tan A = \dfrac{8}{17}$ (3) $\cos A = \dfrac{15}{17}$ (4) $\tan A = \dfrac{5}{8}$

6. In △*ABC*, the measure of ∠*B* = 90°, *AC* = 50, *AB* = 48, and *BC* = 14. Which ratio represents the tangent of ∠*A*?

(1) $\dfrac{14}{50}$ (2) $\dfrac{14}{48}$ (3) $\dfrac{48}{50}$ (4) $\dfrac{48}{14}$

Show Work:

1. A stake is to be driven into the ground away from the base of a 50-foot pole, as shown in the diagram below. A wire from the stake on the ground to the top of the pole is to be installed at an angle of elevation of 52°.

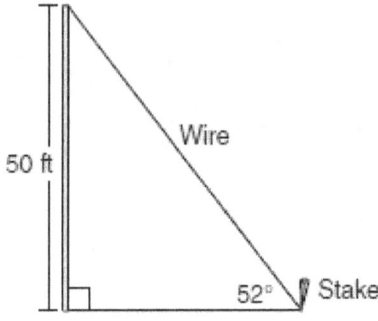

How far away from the base of the pole should the stake be driven in, to the *nearest foot*? What will be the length of the wire from the stake to the top of the pole, to the *nearest foot*?

2. In right triangle ABC, $AB = 20$, $AC = 12$, $BC = 16$, and $m\angle C = 90$. Find, to the *nearest degree*, the measure of $\angle A$.

3. A communications company is building a 30-foot antenna to carry cell phone transmissions. As shown in the diagram below, a 50-foot wire from the top of the antenna to the ground is used to stabilize the antenna.

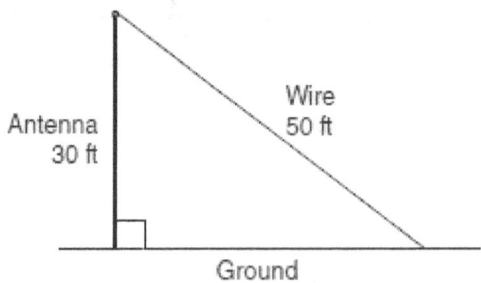

Find, to the *nearest degree*, the measure of the angle that the wire makes with the ground.

1. Luis is going to paint a basketball court on his driveway, as shown in the diagram below. This basketball court consists of a rectangle and a semicircle.

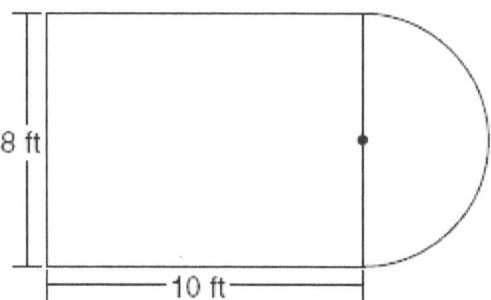

Which expression represents the area of this basketball court, in square feet?
(1) 80
(2) 80 + 8π
(3) 80 + 16π
(4) 80 + 64π

2. Lenny made a cube in technology class. Each edge measured 1.5 cm. What is the volume of the cube in cubic centimeters?
(1) 2.25   (2) 3.375   (3) 9.0   (4) 13.5

3. How many square inches of wrapping paper are needed to entirely cover a box that is 2 inches by 3 inches by 4 inches?
(1) 18   (2) 24   (3) 26   (4) 52

4. A cylindrical container has a diameter of 12 inches and a height of 15 inches, as illustrated in the diagram below.

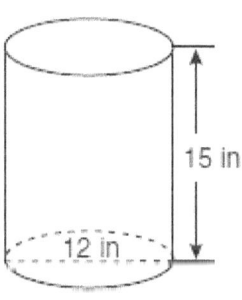

(Not drawn to scale)

What is the volume of this container to the *nearest tenth* of a cubic inch?
(1) 6,785.8   (2) 4,241.2   (3) 2,160.0   (4) 1,696.5

VI. Geometric Measurements

5. The groundskeeper is replacing the turf on a football field. His measurements of the field are 130 yards by 60 yards. The actual measurements are 120 yards by 54 yards. Which expression represents the relative error in the measurement?

(1) $\dfrac{(130)(60) - (120)(54)}{(120)(54)}$

(3) $\dfrac{(130)(60) - (120)(54)}{(130)(60)}$

(2) $\dfrac{(120)(54)}{(130)(60) - (120)(54)}$

(4) $\dfrac{(130)(60)}{(130)(60) - (120)(54)}$

6. To calculate the volume of a small wooden cube, Ezra measured an edge of the cube as 2 cm. The actual length of the edge of Ezra's cube is 2.1 cm. What is the relative error in his volume calculation to the *nearest hundredth*?

(1)   0.13   (2)   0.14   (3)   0.15   (4)   0.16

Show Work:

1. Serena's garden is a rectangle joined with a semicircle, as shown in the diagram below. Line segment *AB* is the diameter of semicircle *P*. Serena wants to put a fence around her garden.

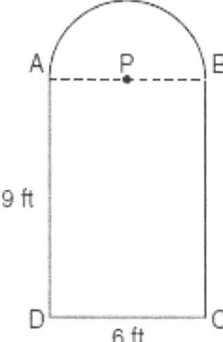

Calculate the length of fence Serena needs to the *nearest tenth of a foot.*

2. A designer created the logo shown below. The logo consists of a square and four quarter-circles of equal size.

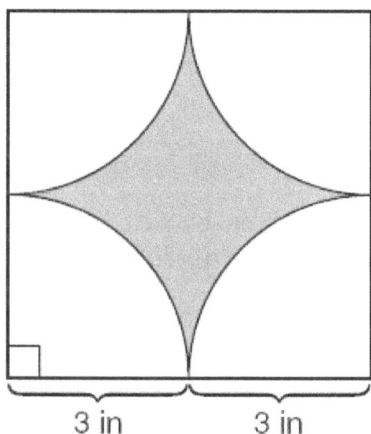

Express, in terms of π, the exact area, in square inches, of the shaded region.

3. In the diagram below, the circumference of circle *O* is 16π inches. The length of *BC* is three-quarters of the length of diameter $\overline{AD}$ and *CE* = 4 inches. Calculate the area, in square inches, of trapezoid *ABCD*.

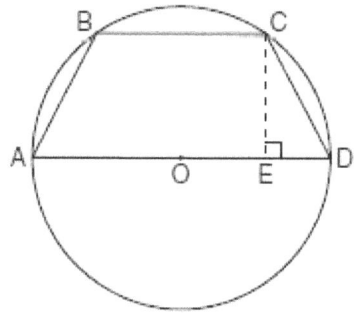

4. A soup can is in the shape of a cylinder. The can has a volume of 342 cm^3 and a diameter of 6 cm. Express the height of the can in terms of π. Determine the maximum number of soup cans that can be stacked on their base between two shelves if the distance between the shelves is exactly 36 cm. Explain your answer.

5. Sophie measured a piece of paper to be 21.7 cm by 28.5 cm. The piece of paper is actually 21.6 cm by 28.4 cm. Determine the number of square centimeters in the area of the piece of paper using Sophie's measurements. Determine the number of square centimeters in the actual area of the piece of paper. Determine the relative error in calculating the area. Express your answer as a decimal to the *nearest thousandth*. Sophie does not think there is a significant amount of error. Do you agree or disagree? Justify your answer.

6. Using his ruler, Howell measured the sides of a rectangular prism to be 5 cm by 8 cm by 4 cm. The actual measurements are 5.3 cm by 8.2 cm by 4.1 cm. Find Howell's relative error in calculating the volume of the prism, to the *nearest thousandth*.

7. Alexis calculates the surface area of a gift box as 600 square inches. The actual surface area of the gift box is 592 square inches. Find the relative error of Alexis' calculation expressed as a decimal to the *nearest thousandth*.

1. What is the slope of the line containing the points $(3, 4)$ and $(-6, 10)$?

(1) $\dfrac{1}{2}$ (2) 2 (3) $-\dfrac{2}{3}$ (4) $-\dfrac{3}{2}$

2. What is an equation of the line that passes through the points $(3, -3)$ and $(-3, -3)$?
(1) $y = 3$ (3) $y = -3$
(2) $x = -3$ (4) $x = y$

3. Which equation represents a line that is parallel to the line $y = 3 - 2x$?
(1) $4x + 2y = 5$ (3) $y = 3 - 4x$
(2) $2x + 4y = 1$ (4) $y = 4x - 2$

4. What is an equation of the line that passes through the point $(4, -6)$ and has a slope of -3?
(1) $y = -3x + 6$ (3) $y = -3x + 10$
(2) $y = -3x - 6$ (4) $y = -3x + 14$

5. Which equation represents the line that passes through the points $(-3, 7)$ and $(3, 3)$?
(1) $y = \dfrac{2}{3}x + 1$ (3) $y = -\dfrac{2}{3}x + 5$
(2) $y = \dfrac{2}{3}x + 9$ (4) $y = -\dfrac{2}{3}x + 9$

6. Which point is on the line $4y - 2x = 0$?
(1) $(-2, -1)$ (2) $(-2, 1)$ (3) $(-1, -2)$ (4) $(1, 2)$

7. Which equation represents a line parallel to the graph of $2x - 4y = 16$?
(1) $y = \dfrac{1}{2}x - 5$ (3) $y = -2x + 6$
(2) $y = -\dfrac{1}{2}x + 4$ (4) $y = 2x + 8$

8. Which linear equation represents a line containing the point $(1, 3)$?
(1) $x + 2y = 5$ (3) $2x + y = 5$
(2) $x - 2y = 5$ (4) $2x - y = 5$

9. The graphs of the equations $y = 2x - 7$ and $y - kx = 7$ are parallel when k equals
(1) -2   (2) 2   (3) -7   (4) 7

10. It takes Tammy 45 minutes to ride her bike 5 miles. At this rate, how long will it take her to ride 8 miles?
(1) 0.89 hour   (2) 1.125 hours   (3) 48 minutes   (4) 72 minutes

11. Consider the graph of the equation $y = ax^2 + bx + c$, when $a \neq 0$. If a is multiplied by 3, what is true of the graph of the resulting parabola?
(1) The vertex is 3 units above the vertex of the original parabola.
(2) The new parabola is 3 units to the right of the original parabola.
(3) The new parabola is wider than the original parabola.
(4) The new parabola is narrower than the original parabola.

12. What are the vertex and axis of symmetry of the parabola $y = x^2 - 16x + 63$?
(1) vertex: $(8, -1)$; axis of symmetry: $x = 8$ (3) vertex: $(-8, -1)$; axis of symmetry: $x = -8$
(2) vertex: $(8, 1)$; axis of symmetry: $x = 8$ (4) vertex: $(-8, 1)$; axis of symmetry: $x = -8$

13. The equation $y = x^2 + 3x - 18$ is graphed on the set of axes below.

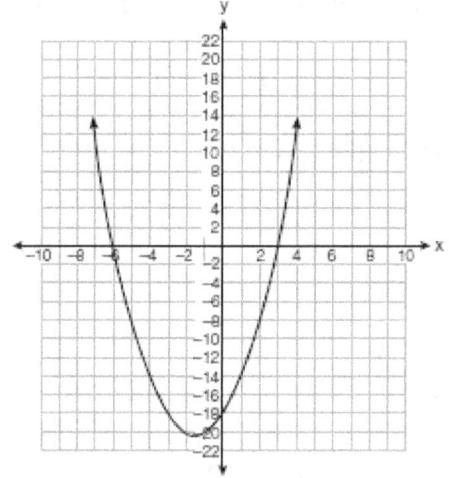

Based on this graph, what are the roots of the equation $x^2 + 3x - 18 = 0$?
(1) −3 and 6   (3) 3 and −6
(2) 0 and −18   (4) 3 and −18

14. What is the equation of the axis of symmetry of the parabola shown in the diagram below?

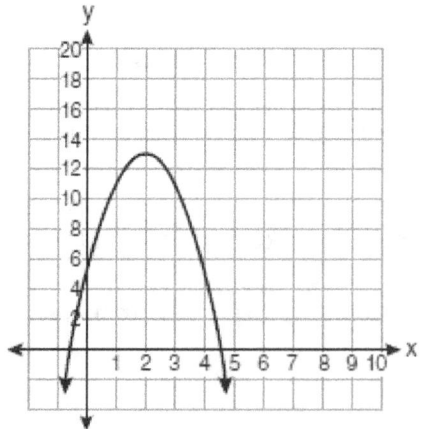

(1) $x = -0.5$ (2) $x = 2$ (3) $x = 4.5$ (4) $x = 13$

15. What are the vertex and axis of symmetry of the parabola shown in the diagram below?

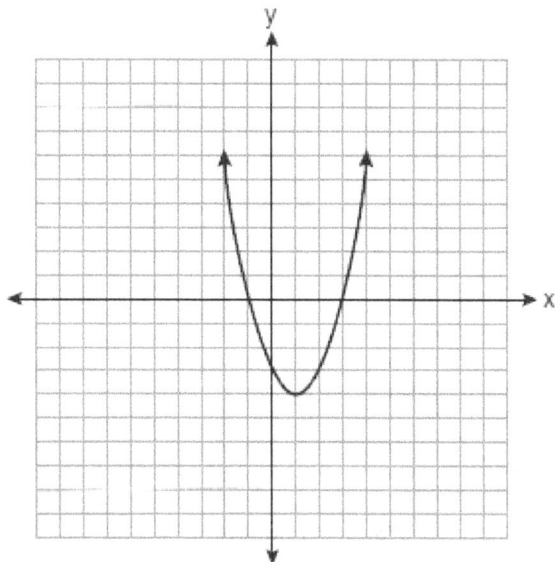

(1) vertex: $(1, -4)$; axis of symmetry: $x = 1$ (3) vertex: $(-4, 1)$; axis of symmetry: $x = 1$
(2) vertex: $(1, -4)$; axis of symmetry: $x = -4$ (4) vertex: $(-4, 1)$; axis of symmetry: $x = -4$

16. Daniel's Print Shop purchased a new printer for $35,000. Each year it depreciates (loses value) at a rate of 5%. What will its approximate value be at the end of the fourth year?
(1) $33,250.00   (3) $28,507.72
(2) $30,008.13   (4) $27,082.33

17. Kathy plans to purchase a car that depreciates (loses value) at a rate of 14% per year. The initial cost of the car is $21,000. Which equation represents the value, v, of the car after 3 years?
(1) $v = 21,000(0.14)^3$ (3) $v = 21,000(1.14)^3$
(2) $v = 21,000(0.86)^3$ (4) $v = 21,000(0.86)(3)$

18. The value, y, of a $15,000 investment over x years is represented by the equation $y = 15000(1.2)^{\frac{x}{3}}$.
What is the profit (interest) on a 6-year investment?
(1) $6,600   (3) $21,600
(2) $10,799   (4) $25,799

19. In a science fiction novel, the main character found a mysterious rock that decreased in size each day. The table below shows the part of the rock that remained at noon on successive days.

Day	Fractional Part of the Rock Remaining
1	1
2	$\frac{1}{2}$
3	$\frac{1}{4}$
4	$\frac{1}{8}$

Which fractional part of the rock will remain at noon on day 7?

(1) $\frac{1}{128}$ (2) $\frac{1}{64}$ (3) $\frac{1}{14}$ (4) $\frac{1}{12}$

20. Which ordered pair is a solution of the system of equations shown in the graph below?

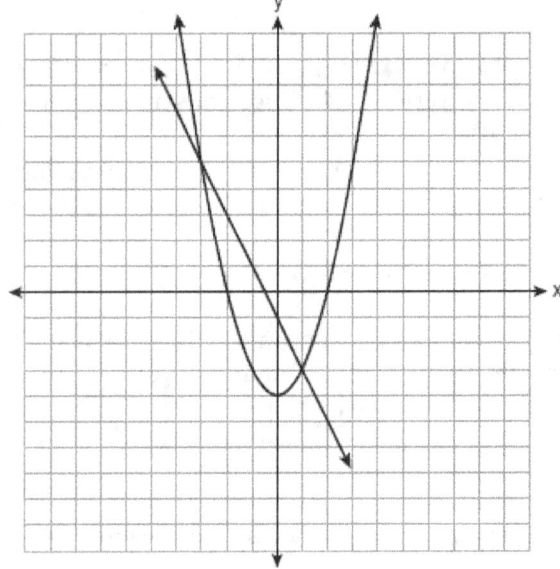

(1) $(-3, 1)$ (2) $(-3, 5)$ (3) $(0, -1)$ (4) $(0, -4)$

21. Which graph represents the solution of $3y - 9 \leq 6x$?

(1)

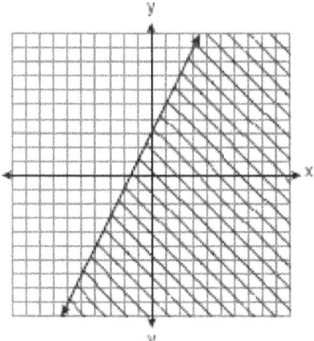

(3)

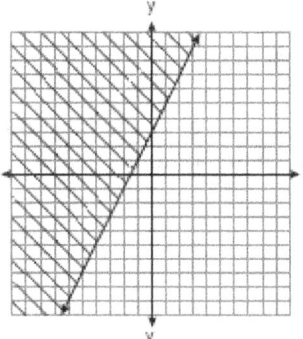

(2)

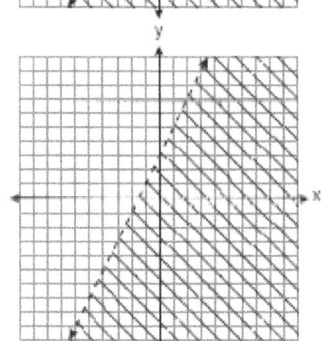

(4)

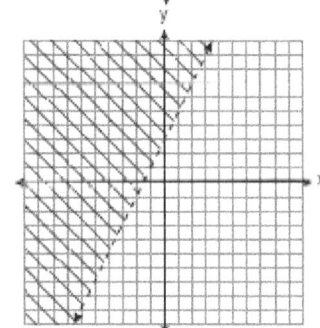

22. Which quadrant will be completely shaded in the graph of the inequality $y \leq 2x$?

(1) Quadrant I   (3) Quadrant III
(2) Quadrant II   (4) Quadrant IV

23. Which ordered pair is in the solution set of the system of linear inequalities graphed below?

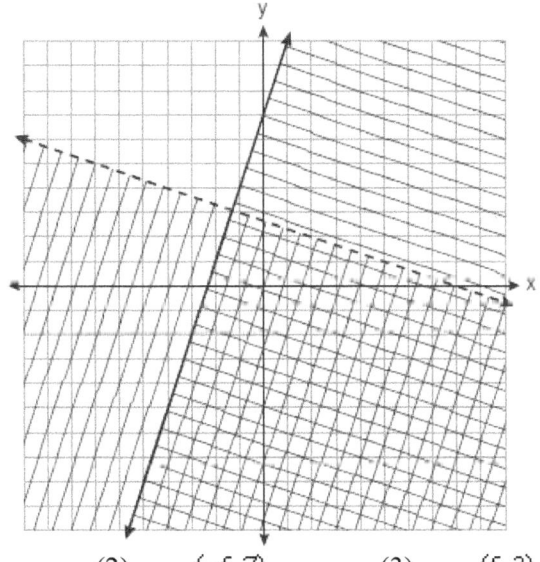

(1) $(1, -4)$ (2) $(-5, 7)$ (3) $(5, 3)$ (4) $(-7, -2)$

24. Which ordered pair is in the solution set of the following system of linear inequalities?

$$y < 2x + 2$$

$$y \geq -x - 1$$

(1) $(0, 3)$ (2) $(2, 0)$ (3) $(-1, 0)$ (4) $(-1, -4)$

25. Which relation represents a function?

(1) $\{(0,3),(2,4),(0,6)\}$

(3) $\{(2,0),(6,2),(6,-2)\}$

(2) $\{(-7,5),(-7,1),(-10,3),(-4,3)\}$

(4) $\{(-6,5),(-3,2),(1,2),(6,5)\}$

26. Antwaan leaves a cup of hot chocolate on the counter in his kitchen. Which graph is the best representation of the change in temperature of his hot chocolate over time?

(1)

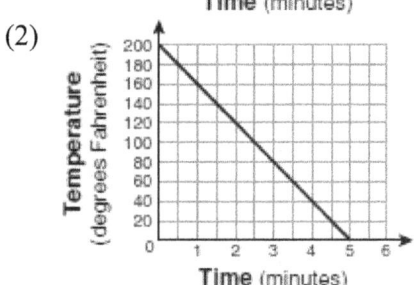

(3)

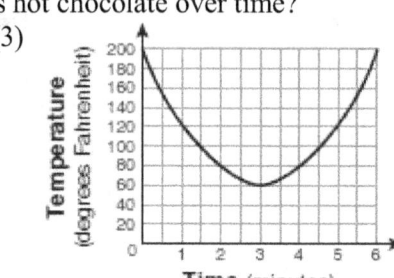

(2)

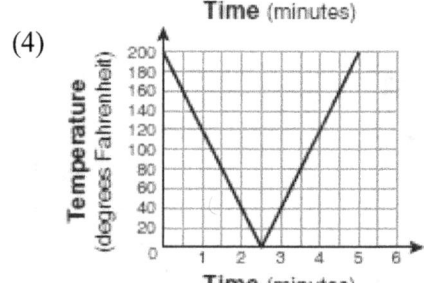

(4)

27. Which graph represents a function?

(1)

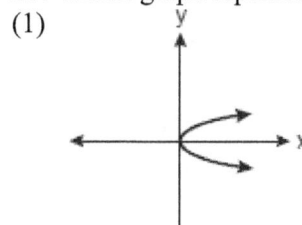

(3)

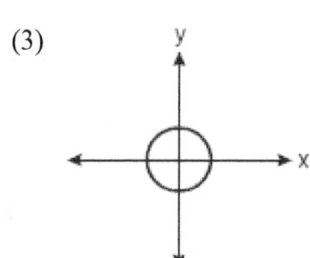

(2)

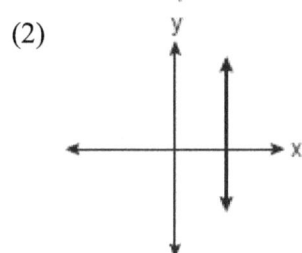

(4)

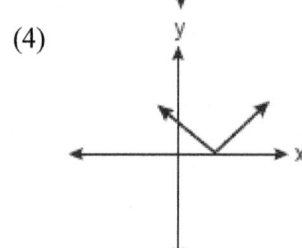

Show Work:

1. Graph and label the following equations on the set of axes below.

$$y = |x|$$

$$y = \left|\frac{1}{2}x\right|$$

Explain how *decreasing* the coefficient of x affects the graph of the equation $y = |x|$.

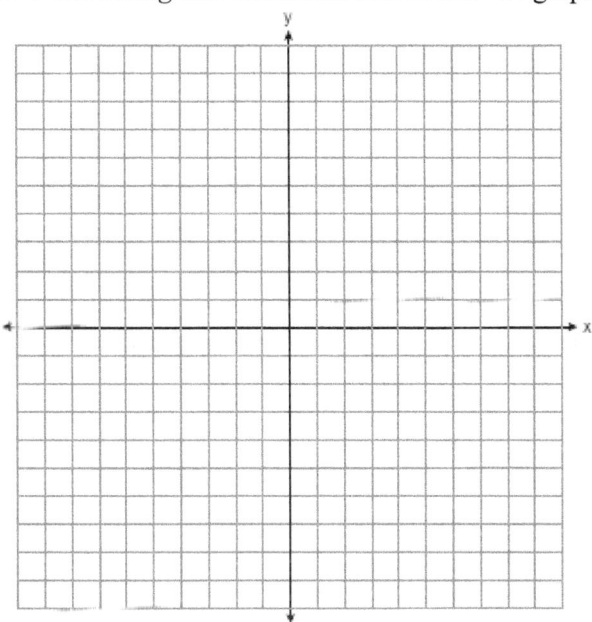

2. Graph the equation $y = x^2 - 2x - 3$ on the accompanying set of axes. Using the graph, determine the roots of the equation $x^2 - 2x - 3 = 0$.

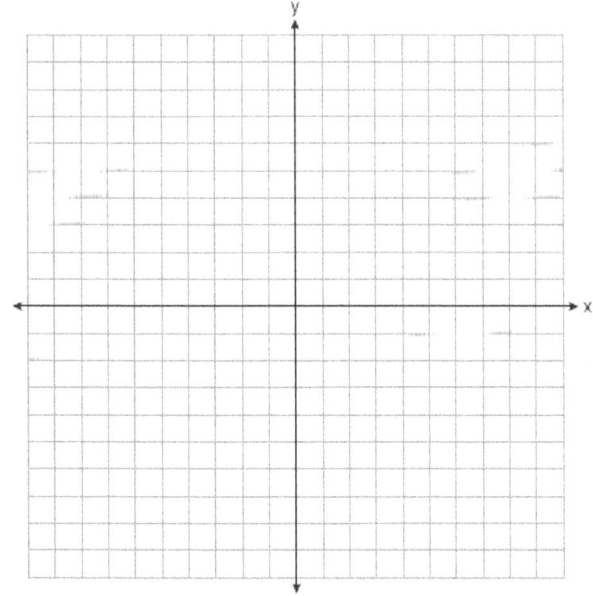

3. A bank is advertising that new customers can open a savings account with a $3\frac{3}{4}\%$ interest rate compounded annually. Robert invests $5,000 in an account at this rate. If he makes no additional deposits or withdrawals on his account, find the amount of money he will have, to the *nearest cent*, after three years.

4. Solve the following systems of equations graphically, on the set of axes below, and state the coordinates of the point(s) in the solution set.

$$y = x^2 - 6x + 5$$

$$2x + y = 5$$

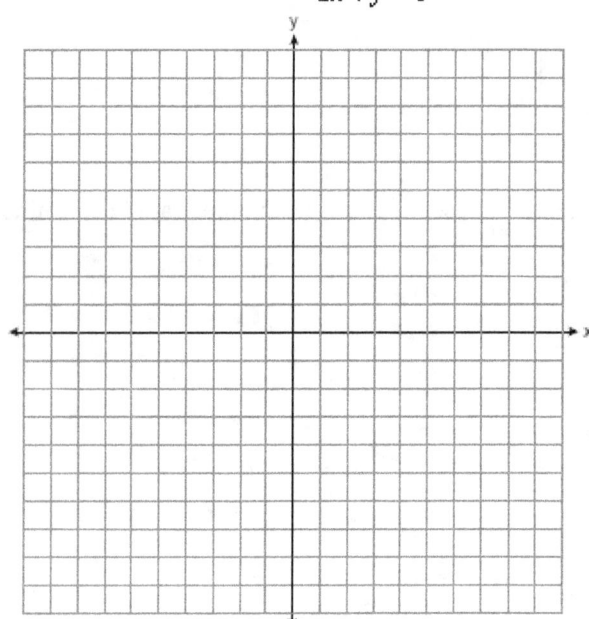

5. Graph the solution set for the inequality $4x - 3y > 9$ on the set of axes below. Determine if the point $(1, -3)$ is in the solution set. Justify your answer.

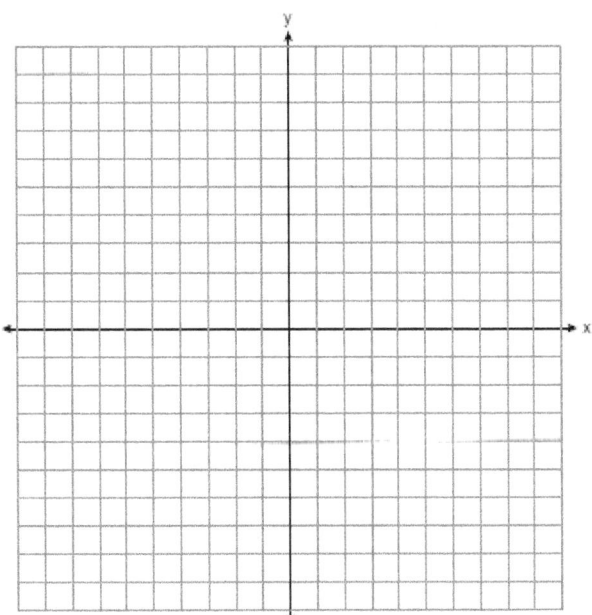

6. On the set of axes below, graph the following system of inequalities and state the coordinates of a point in the solution set.

$$2x - y \geq 6$$

$$x > 2$$

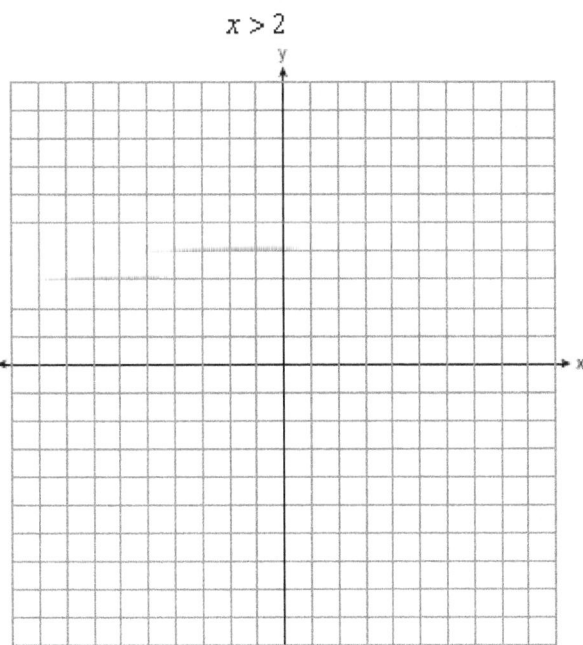

1. A spinner is divided into eight equal regions as shown in the diagram below.

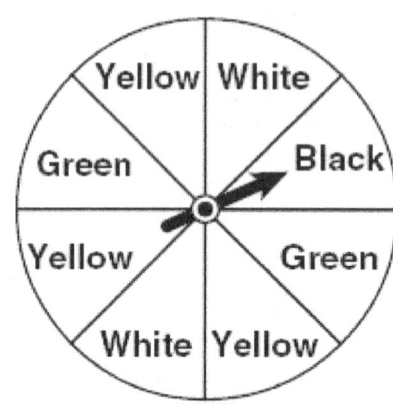

Which event is most likely to occur in one spin?
(1) The arrow will land in a green or white area.
(2) The arrow will land in a green or black area.
(3) The arrow will land in a yellow or black area.
(4) The arrow will land in a yellow or green area.

2. The faces of a cube are numbered from 1 to 6. If the cube is tossed once, what is the probability that a prime number or a number divisible by 2 is obtained?

(1) $\dfrac{6}{6}$ (2) $\dfrac{5}{6}$ (3) $\dfrac{4}{6}$ (4) $\dfrac{1}{6}$

3. The faces of a cube are numbered from 1 to 6. If the cube is rolled once, which outcome is *least* likely to occur?
(1) rolling an odd number   (3) rolling a number less than 6
(2) rolling an even number   (4) rolling a number greater than 4

4. Students in Ms. Nazzeer's mathematics class tossed a six-sided number cube whose faces are numbered 1 to 6. The results are recorded in the table below.

Result	Frequency
1	3
2	6
3	4
4	6
5	4
6	7

Based on these data, what is the empirical probability of tossing a 4?

(1) $\dfrac{8}{30}$ (2) $\dfrac{6}{30}$ (3) $\dfrac{5}{30}$ (4) $\dfrac{1}{30}$

5. A bag contains eight green marbles, five white marbles, and two red marbles. What is the probability of drawing a red marble from the bag?

(1) $\frac{1}{15}$ (2) $\frac{2}{15}$ (3) $\frac{2}{13}$ (4) $\frac{13}{15}$

6. Three high school juniors, Reese, Matthew, and Chris, are running for student council president. A survey is taken a week before the election asking 40 students which candidate they will vote for in the election. The results are shown in the table below.

Candidate's Name	Number of Students Supporting Candidate
Reese	15
Matthew	13
Chris	12

Based on the table, what is the probability that a student will vote for Reese?

(1) $\frac{1}{3}$ (2) $\frac{3}{5}$ (3) $\frac{3}{8}$ (4) $\frac{5}{8}$

7. Keisha is playing a game using a wheel divided into eight equal sectors, as shown in the diagram below. Each time the spinner lands on orange, she will win a prize.

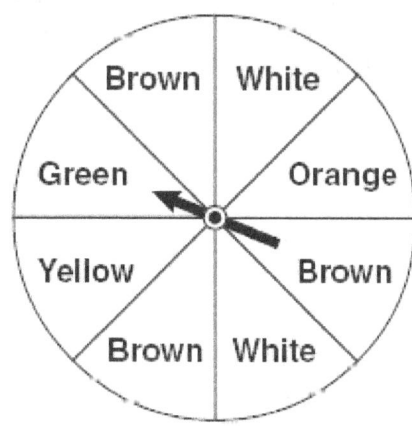

If Keisha spins this wheel twice, what is the probability she will win a prize on *both* spins?

(1) $\frac{1}{64}$ (2) $\frac{1}{56}$ (3) $\frac{1}{16}$ (4) $\frac{1}{4}$

8. How many different sandwiches consisting of one type of cheese, one condiment, and one bread choice can be prepared from five types of cheese, two condiments, and three bread choices?
(1) 10   (2) 13   (3) 15   (4) 30

9. John is going to line up his four golf trophies on a shelf in his bedroom. How many different possible arrangements can he make?
(1) 24   (2) 16   (3) 10   (4) 4

10. The bowling team at Lincoln High School must choose a president, vice president, and secretary. If the team has 10 members, which expression could be used to determine the number of ways the officers could be chosen?

(1) $_3P_{10}$ (2) $_7P_3$ (3) $_{10}P_3$ (4) $_{10}P_7$

11. How many different three-letter arrangements can be formed using the letters in the word *ABSOLUTE* if each letter is used only once?

(1)   56   (2)   112   (3)   168   (4)   336

Show Work:

1. Some books are laid on a desk. Two are English, three are mathematics, one is French, and four are social studies. Theresa selects an English book and Isabelle then selects a social studies book. Both girls take their selections to the library to read. If Truman then selects a book at random, what is the probability that he selects an English book?

2. Jon is buying tickets for himself for two concerts. For the jazz concert, 4 tickets are available in the front row, and 32 tickets are available in the other rows. For the orchestra concert, 3 tickets are available in the front row, and 23 tickets are available in the other rows. Jon is randomly assigned one ticket for each concert. Determine the concert for which he is more likely to get a front-row ticket. Justify your answer.

3. A password consists of three digits, 0 through 9, followed by three letters from an alphabet having 26 letters. If repetition of digits is allowed, but repetition of letters is not allowed, determine the number of different passwords that can be made. If repetition is not allowed for digits or letters, determine how many fewer different passwords can be made.

4. A restaurant sells kids' meals consisting of one main course, one side dish, and one drink, as shown in the table below.

Kids' Meal Choices

Main Course	Side Dish	Drink
hamburger	French fries	milk
chicken nuggets	applesauce	juice
turkey sandwich		soda

Draw a tree diagram or list the sample space showing all possible kids' meals. How many different kids' meals can a person order? Jose does not drink juice. Determine the number of different kids' meals that do *not* include juice. Jose's sister will eat *only* chicken nuggets for her main course. Determine the number of different kids' meals that include chicken nuggets.

1. A school wants to add a coed soccer program. To determine student interest in the program, a survey will be taken. In order to get an unbiased sample, which group should the school survey?
(1) every third student entering the building
(2) every member of the varsity football team
(3) every member in Ms. Zimmer's drama classes
(4) every student having a second-period French class

2. Which data set describes a situation that could be classified as qualitative?
(1) the elevations of the five highest mountains in the world
(2) the ages of presidents at the time of their inauguration
(3) the opinions of students regarding school lunches
(4) the shoe sizes of players on the basketball team

3. Alex earned scores of 60, 74, 82, 87, 87, and 94 on his first six algebra tests. What is the relationship between the measures of central tendency of these scores?
(1) median < mode < mean   (3) mode < median < mean
(2) mean < mode < median   (4) mean < median < mode

4. A survey is being conducted to determine which types of television programs people watch. Which survey and location combination would likely contain the most bias?
(1) surveying 10 people who work in a sporting goods store
(2) surveying the first 25 people who enter a grocery store
(3) randomly surveying 50 people during the day in a mall
(4) randomly surveying 75 people during the day in a clothing store

5. Which data set describes a situation that could be classified as qualitative?
(1) the ages of the students in Ms. Marshall's Spanish class
(2) the test scores of the students in Ms. Fitzgerald's class
(3) the favorite ice cream flavor of each of Mr. Hayden's students
(4) the heights of the players on the East High School basketball team

6. The freshman class held a canned food drive for 12 weeks. The results are summarized in the table below.

Canned Food Drive Results

Week	1	2	3	4	5	6	7	8	9	10	11	12
Number of Cans	20	35	32	45	58	46	28	23	31	79	65	62

Which number represents the second quartile of the number of cans of food collected?
(1)   29.5   (2)   30.5   (3)   40   (4)   60

7. Which data table represents univariate data?

(1)

Side Length of a Square	Area of Square
2	4
3	9
4	16
5	25

(2)

Hours Worked	Pay
20	$160
25	$200
30	$240
35	$280

(3)

Age Group	Frequency
20–29	9
30–39	7
40–49	10
50–59	4

(4)

People	Number of Fingers
2	20
3	30
4	40
5	50

8. Which table does *not* show bivariate data?

(1)

Height (inches)	Weight (pounds)
39	50
48	70
60	90

(2)

Gallons	Miles Driven
15	300
20	400
25	500

(3)

Quiz Average	Frequency
70	12
80	15
90	6

(4)

Speed (mph)	Distance (miles)
40	80
50	120
55	150

9. Four hundred licensed drivers participated in the math club's survey on driving habits. The table below shows the number of drivers surveyed in each age group.

**Ages of People in Survey on
Driving Habits**

Age Group	Number of Drivers
16–25	150
26–35	129
36–45	33
46–55	57
56–65	31

Which statement best describes a conclusion based on the data in the table?

(1) It may be biased because no one younger than 16 was surveyed.

(2) It would be fair because many different age groups were surveyed.

(3) It would be fair because the survey was conducted by the math club students.

(4) It may be biased because the majority of drivers surveyed were in the younger age intervals.

10. The box-and-whisker plot below represents the math test scores of 20 students.

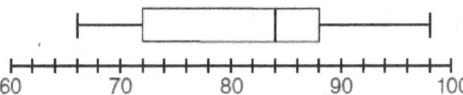

What percentage of the test scores are *less than* 72?

(1) 25   (2) 50   (3) 75   (4) 100

11. The data set 5, 6, 7, 8, 9, 9, 9, 10, 12, 14, 17, 17, 18, 19, 19 represents the number of hours spent on the Internet in a week by students in a mathematics class. Which box-and-whisker plot represents the data?

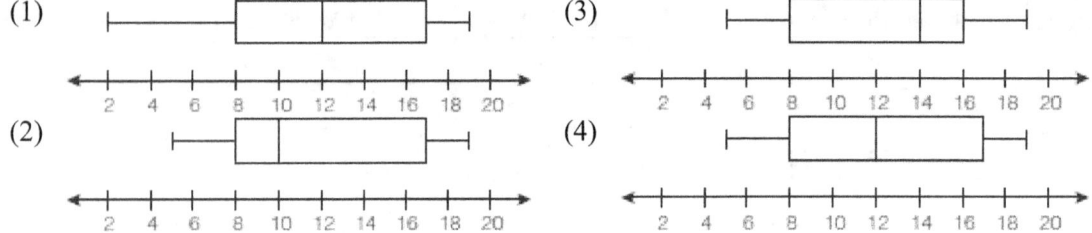

12. The box-and-whisker plot below represents students' scores on a recent English test.

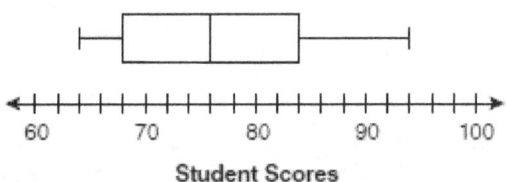

Student Scores

What is the value of the upper quartile?

(1) 68   (2) 76   (3) 84   (4) 94

13. A movie theater recorded the number of tickets sold daily for a popular movie during the month of June. The box-and-whisker plot shown below represents the data for the number of tickets sold, in hundreds.

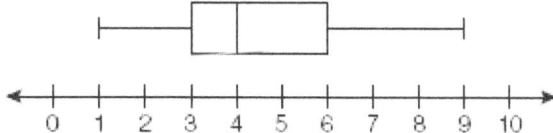

Which conclusion can be made using this plot?
(1) The second quartile is 600.
(2) The mean of the attendance is 400.
(3) The range of the attendance is 300 to 600.
(4) Twenty-five percent of the attendance is between 300 and 400.

14. The table below shows a cumulative frequency distribution of runners' ages.

Cumulative Frequency Distribution of Runners' Ages

Age Group	Total
20–29	8
20–39	18
20–49	25
20–59	31
20–69	35

According to the table, how many runners are in their forties?
(1)   25   (2)   10   (3)   7   (4)   6

15. Which situation describes a correlation that is *not* a causal relationship?
(1) The rooster crows, and the Sun rises.
(2) The more miles driven, the more gasoline needed
(3) The more powerful the microwave, the faster the food cooks.
(4) The faster the pace of a runner, the quicker the runner finishes.

16. Which situation should be analyzed using bivariate data?
(1) Ms. Saleem keeps a list of the amount of time her daughter spends on her social studies homework.
(2) Mr. Benjamin tries to see if his students' shoe sizes are directly related to their heights.
(3) Mr. DeStefan records his customers' best video game scores during the summer.
(4) Mr. Chan keeps track of his daughter's algebra grades for the quarter.

17. Which situation describes a correlation that is *not* a causal relationship?
(1) the length of the edge of a cube and the volume of the cube
(2) the distance traveled and the time spent driving
(3) the age of a child and the number of siblings the child has
(4) the number of classes taught in a school and the number of teachers employed

18. There is a negative correlation between the number of hours a student watches television and his or her social studies test score. Which scatter plot below displays this correlation?

(1)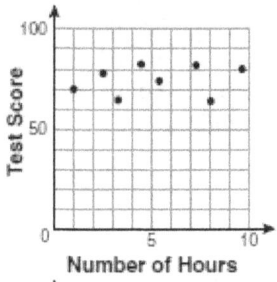

(3)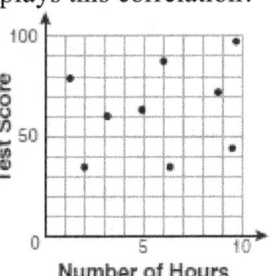

(2)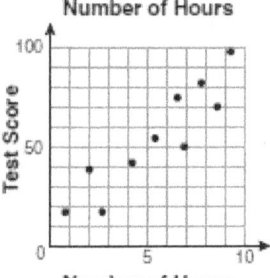

(4)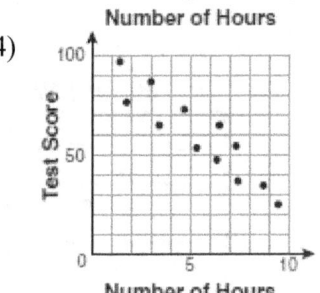

19. Which scatter plot shows the relationship between *x* and *y* if *x* represents a student score on a test and *y* represents the number of incorrect answers a student received on the same test?

(1)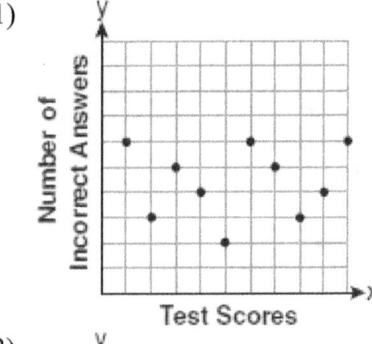

(3)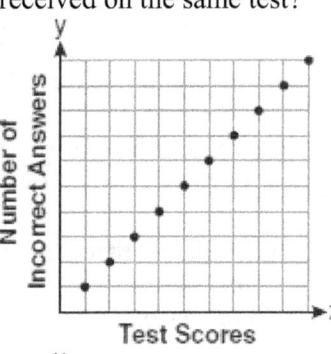

(2)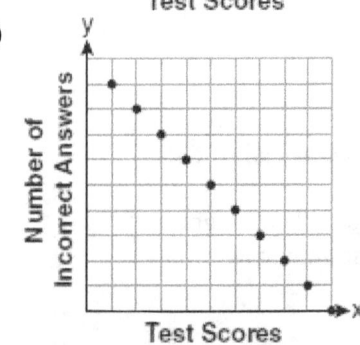

(4)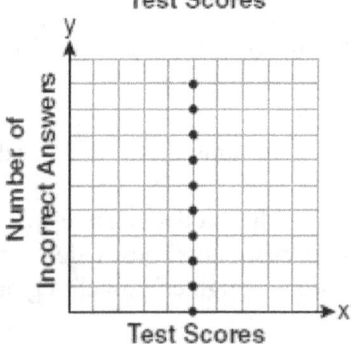

Show Work:

1. The values of 11 houses on Washington St. are shown in the table below.

Value per House	Number of Houses
$100,000	1
$175,000	5
$200,000	4
$700,000	1

Find the mean value of these houses in dollars. Find the median value of these houses in dollars. State which measure of central tendency, the mean or the median, *best* represents the values of these 11 houses. Justify your answer.

2. Ms. Mosher recorded the math test scores of six students in the table below.

Student	Student Score
Andrew	72
John	80
George	85
Amber	93
Betty	78
Roberto	80

Determine the mean of the student scores, to the *nearest tenth*. Determine the median of the student scores. Describe the effect on the mean and the median if Ms. Mosher adds 5 bonus points to each of the six students' scores.

3. The Fahrenheit temperature readings on 30 April mornings in Stormville, New York, are shown below.

41°, 58°, 61°, 54°, 49°, 46°, 52°, 58°, 67°, 43°, 47°, 60°, 52°, 58°, 48°,
44°, 59°, 66°, 62°, 55°, 44°, 49°, 62°, 61°, 59°, 54°, 57°, 58°, 63°, 60°

Using the data, complete the frequency table below.

Interval	Tally	Frequency
40–44		
45–49		
50–54		
55–59		
60–64		
65–69		

On the grid below, construct and label a frequency histogram based on the table.

4. The diagram below shows a cumulative frequency histogram of the students' test scores in Ms. Wedow's algebra class.

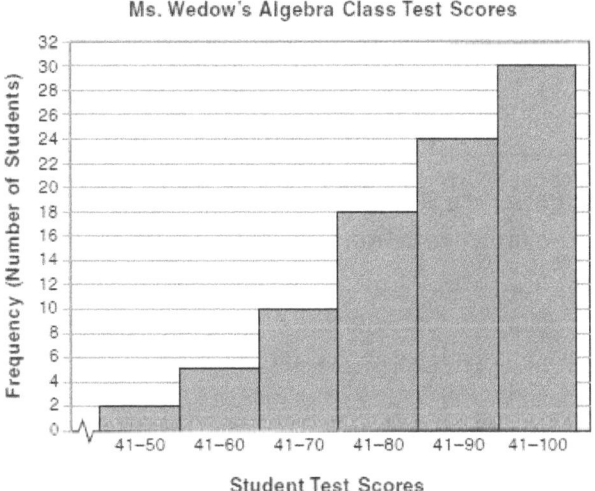

Determine the total number of students in the class. Determine how many students scored higher than 70. State which *ten-point interval* contains the median. State which *two ten-point* intervals contain the same frequency.

5. Megan and Bryce opened a new store called the Donut Pit. Their goal is to reach a profit of $20,000 in their 18th month of business. The table and scatter plot below represent the profit, P, in thousands of dollars, that they made during the first 12 months.

t (months)	P (profit, in thousands of dollars)
1	3.0
2	2.5
3	4.0
4	5.0
5	6.5
6	5.5
7	7.0
8	6.0
9	7.5
10	7.0
11	9.0
12	9.5

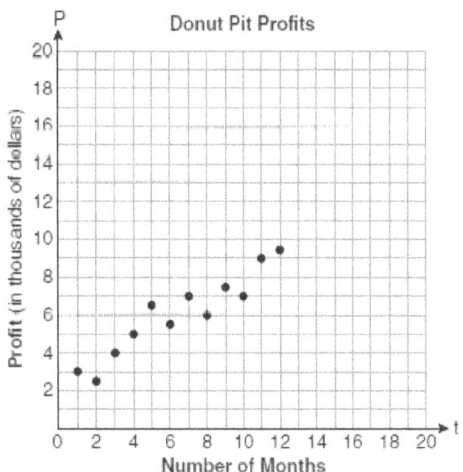

Draw a reasonable line of best fit. Using the line of best fit, predict whether Megan and Bryce will reach their goal in the 18th month of their business. Justify your answer.

6. The table below shows the number of prom tickets sold over a ten-day period.

Prom Ticket Sales

Day (x)	1	2	5	7	10
Number of Prom Tickets Sold (y)	30	35	55	60	70

Plot these data points on the coordinate grid below. Use a consistent and appropriate scale. Draw a reasonable line of best fit and write its equation.

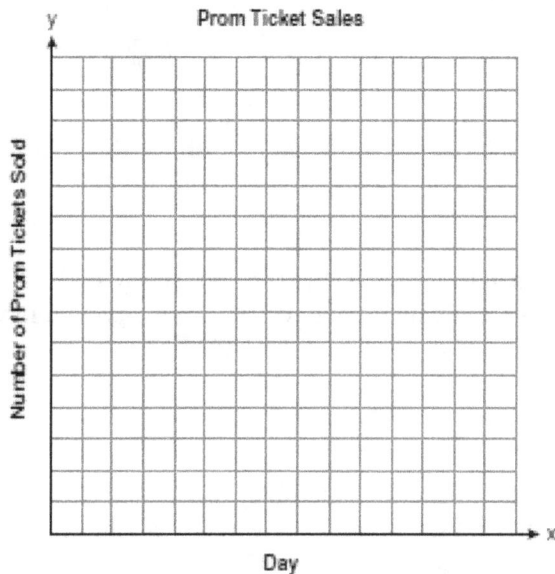

1. Which interval notation represents the set of all numbers from 2 through 7, inclusive?
   (1)   (2, 7]   (2)   (2, 7)   (3)   [2, 7)   *(4)   **[2, 7]**

The interval notation [2, 7] represnts $2 \le x \le 7$.

---

2. The set $\{1, 2, 3, 4\}$ is equivalent to
   (1) $\left\{ x \mid 1 < x < 4, \text{ where } x \text{ is a whole number} \right\}$ *(3) $\left\{ x \mid 0 < x \le 4, \text{ where } x \text{ is a whole number} \right\}$

   (2) $\left\{ x \mid 0 < x < 4, \text{ where } x \text{ is a whole number} \right\}$ (4) $\left\{ x \mid 1 < x \le 4, \text{ where } x \text{ is a whole number} \right\}$

3. The set $\{11, 12\}$ is equivalent to
   (1) $\{ x \mid 11 < x < 12, \text{ where } x \text{ is an integer} \}$ (3) $\{ x \mid 10 \le x < 12, \text{ where } x \text{ is an integer} \}$
   (2) $\{ x \mid 11 < x \le 12, \text{ where } x \text{ is an integer} \}$ *(4) $\{ x \mid 10 < x \le 12, \text{ where } x \text{ is an integer} \}$

4. Which interval notation represents the set of all numbers greater than or equal to 5 and less than 12?
   *(1)   [5, 12)   (2)   (5, 12]   (3)   (5, 12)   (4)   [5, 12]

The interval notation [5, 12) represnts $5 \le x < 12$.

---

5. Given: Set $A = \{(-2, -1), (-1, 0), (1, 8)\}$

            Set $B = \{(-3, -4), (-2, -1), (-1, 2), (1, 8)\}$.

   What is the intersection of sets A and B?
   (1) $\{(1, 8)\}$ *(3) $\{(-2, -1), (1, 8)\}$
   (2) $\{(-2, -1)\}$ (4) $\{(-3, -4), (-2, -1), (-1, 2), (-1, 0), (1, 8)\}$

All elements of the intersection belong to both sets A and B.

---

6. Consider the set of integers greater than -2 and less than 6. A subset of this set is the positive
   factors of 5. What is the complement of this subset?
   (1) $\{0, 2, 3, 4\}$ (3) $\{-2, -1, 0, 2, 3, 4, 6\}$
   *(2) $\{-1, 0, 2, 3, 4\}$ (4) $\{-2, -1, 0, 1, 2, 3, 4, 5, 6\}$

The set is {-1, 0, 1, 2, 3, 4, 5}.
The subset is {1, 5}.
The complement of the subset is {-1, 0, 2, 3, 4}.

---

7. Given: $Q = \{0, 2, 4, 6\}$

   $W = \{0, 1, 2, 3\}$

   $Z = \{1, 2, 3, 4\}$

What is the intersection of sets Q, W, and Z?

*(1) $\{2\}$ (2) $\{0, 2\}$ (3) $\{1, 2, 3\}$ (4) $\{0, 1, 2, 3, 4, 6\}$

Only 2 is the common element of the three sets.

---

8. Given: Set $U = \{S, O, P, H, I, A\}$

   Set $B = \{A, I, O\}$

If set B is a subset of set U, what is the complement of set B?

(1) $\{O, P, S\}$ (2) $\{I, P, S\}$ (3) $\{A, H, P\}$ ***(4)** $\{H, P, S\}$

9. The value of the expression $-|a - b|$ when $a = 7$ and $b = -3$ is

*(1)   **-10**   (2)   10   (3)   -4   (4)   4

$- |a - b| = - |7 - (-3)| = - |7 + 3| = -10$

---

10. Which property is illustrated by the equation $ax + ay = a(x + y)$?

(1)   associative   (2)   commutative   ***(3)   distributive**   (4)   identity

11. What is the additive inverse of the expression $a - b$?

(1) $a + b$ (2) $a - b$ ***(3)** $-a + b$ (4) $-a - b$

$-(a - b) = -a + b$

---

12. Debbie solved the linear equation $3(x + 4) - 2 = 16$ as follows:

   [Line 1] $3(x + 4) - 2 = 16$
   [Line 2] $3(x + 4) = 18$
   [Line 3] $3x + 4 = 18$
   [Line 4] $3x = 14$
   [Line 5] $x = 4\frac{2}{3}$

She made an error between lines

(1)   1 and 2   ***(2)   2 and 3**   (3)   3 and 4   (4)   4 and 5

[Line 2] $3(x + 4) = 18$
[Line 3] $3x + 3 \bullet 4 = 18$ that is $3x + 12 = 18$

Show Work:

1. Maureen tracks the range of outdoor temperatures over three days. She records the following information.

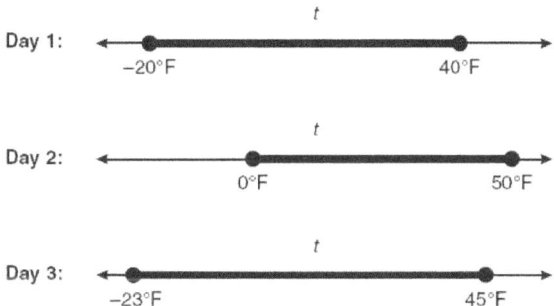

Express the intersection of the three sets as an inequality in terms of temperature, t.

The intersection is the common part.
The greatest lower bound is 0°F ; the least upper bound is 40°F.
$$0 \leq t \leq 40$$

---

2. Perform the indicated operation: $-6(a - 7)$
State the name of the property used.

-6(a - 7) = -6a + 42   Distributive Property

---

3. The Hudson Record Store is having a going-out-of-business sale. CDs normally sell for $18.00. During the first week of the sale, all CDs will sell for $15.00. Written as a fraction, what is the rate of discount? What is this rate expressed as a percent? Round your answer to the *nearest hundredth of a percent*. During the second week of the sale, the same CDs will be on sale for 25% off the *original* price. What is the price of a CD during the second week of the sale?

$$\frac{18 - 15}{18} = \frac{3}{18} = \frac{1}{6}$$
$$\frac{1}{6} \bullet 100\% = 16.67\%$$
18 • (1 - 25%) = 18 • 0.75 = $13.50

60. II. Algebraic Expressions and Operations

1. What is the product of $-3x^2y$ and $(5xy^2 + xy)$?
*(1) $-15x^3y^3 - 3x^3y^2$ \qquad (3) $-15x^2y^2 - 3x^2y$
 (2) $-15x^3y^3 - 3x^3y$ \qquad (4) $-15x^3y^3 + xy$

$-3x^2y\,(5xy^2 + xy) = -3x^2y \bullet 5xy^2 - 3x^2y \bullet xy = -15x^3y^3 - 3x^3y^2$ \hfill Distributive Property

---

2. When $4x^2 + 7x - 5$ is subtracted from $9x^2 - 2x + 3$, the result is
 (1) $5x^2 + 5x - 2$ \qquad (3) $-5x^2 + 5x - 2$
*(2) $5x^2 - 9x + 8$ \qquad (4) $-5x^2 + 9x - 8$

$9x^2 - 2x + 3 - (4x^2 + 7x - 5)$
$= 9x^2 - 2x + 3 - 4x^2 - 7x + 5$ \quad Change the sign of every term in the - (   )
$= 5x^2 - 9x + 8$ \quad Combine like terms

---

3. The sum of $4x^3 + 6x^2 + 2x - 3$ and $3x^3 + 3x^2 - 5x - 5$ is
 (1) $7x^3 + 3x^2 - 3x - 8$ \qquad *(3) $7x^3 + 9x^2 - 3x - 8$
 (2) $7x^3 + 3x^2 + 7x + 2$ \qquad (4) $7x^6 + 9x^4 - 3x^2 - 8$

$4x^3 + 6x^2 + 2x - 3 + 3x^3 + 3x^2 - 5x - 5$
$= 7x^3 + 9x^2 - 3x - 8$ \qquad Combine like terms

---

4. Factored, the expression $16x^2 - 25y^2$ is equivalent to
*(1) $(4x - 5y)(4x + 5y)$ \qquad (3) $(8x - 5y)(8x + 5y)$
 (2) $(4x - 5y)(4x - 5y)$ \qquad (4) $(8x - 5y)(8x - 5y)$

$16x^2 - 25y^2 = (4x)^2 - (5y)^2 = (4x - 5y)(4x + 5y)$

---

5. If Ann correctly factors an expression that is the difference of two perfect squares, her factors could be
 (1) $(2x + y)(x - 2y)$ \qquad (3) $(x - 4)(x - 4)$
*(2) $(2x + 3y)(2x - 3y)$ \qquad (4) $(2y - 5)(y - 5)$

6. Factored completely, the expression $3x^2 - 3x - 18$ is equivalent to
 (1) $3(x^2 - x - 6)$ \qquad (3) $(3x - 9)(x + 2)$
*(2) $3(x - 3)(x + 2)$ \qquad (4) $(3x + 6)(x - 3)$

$3x^2 - 3x - 18 = 3(x^2 - x - 6) = 3(x - 3)(x + 2)$

7. Which expression represents $\dfrac{(2x^3)(8x^5)}{4x^6}$ in simplest form?

(1) x^2 (2) x^9 *(3) $4x^2$ (4) $4x^9$

$$\frac{(2x^3)(8x^5)}{4x^6} = \frac{16x^{3+5}}{4x^6} = \frac{16x^8}{4x^6} = 4x^2$$

8. The expression $\dfrac{9x^4 - 27x^6}{3x^3}$ is equivalent to

(1) $3x(1-3x)$ (3) $3x(1-9x^5)$

*(2) $3x(1-3x^2)$ (4) $9x^3(1-x)$

$$\frac{9x^4 - 27x^6}{3x^3} = \frac{9x^4(1-3x^2)}{3x^3} = 3x(1-3x^2)$$

9. What is the product of $\dfrac{x^2-1}{x+1}$ and $\dfrac{x+3}{3x-3}$ expressed in simplest form?

(1) x (2) $\dfrac{x}{3}$ (3) $x+3$ *(4) $\dfrac{x+3}{3}$

$$\frac{(x^2-1)}{(x+1)} \bullet \frac{(x+3)}{(3x-3)} = \frac{(x+1)(x-1)}{(x+1)} \bullet \frac{(x+3)}{3(x-1)} = \frac{x+3}{3}$$

10. Which expression represents $\dfrac{2x^2 - 12x}{x-6}$ in simplest form?

(1) 0 *(2) $2x$ (3) $4x$ (4) $2x+2$

$$\frac{2x^2 - 12x}{x-6} = \frac{2x(x-6)}{(x-6)} = 2x$$

11. What is $\dfrac{6}{5x} - \dfrac{2}{3x}$ in simplest form?

(1) $\dfrac{8}{15x^2}$ *(2) $\dfrac{8}{15x}$ (3) $\dfrac{4}{15x}$ (4) $\dfrac{4}{2x}$

$$\frac{6}{5x} - \frac{2}{3x} = \frac{6 \cdot 3}{5x \cdot 3} - \frac{2 \cdot 5}{3x \cdot 5} = \frac{18}{15x} - \frac{10}{15x} = \frac{18 - 10}{15x} = \frac{8}{15x}$$

12. The function $y = \dfrac{x}{x^2 - 9}$ is undefined when the value of x is

(1) 0 or 3   *(2) 3 or -3 (3) 3, only   (4) -3, only

$x^2 - 9 = 0$, $(x + 3)(x - 3) = 0$
The function is undefined when x = -3 or x = 3 .

13. Which value of n makes the expression $\dfrac{5n}{2n-1}$ undefined?

(1) 1   (2) 0   (3) $-\dfrac{1}{2}$ *(4) $\dfrac{1}{2}$

$2n - 1 = 0$, $2n = 1$, $n = \dfrac{1}{2}$ the function is undefined.

14. Which expression represents $\dfrac{x^2 - 2x - 15}{x^2 + 3x}$ in simplest form?

(1) -5 *(2) $\dfrac{x-5}{x}$ (3) $\dfrac{-2x-5}{x}$ (4) $\dfrac{-2x-15}{3x}$

$$\frac{x^2 - 2x - 15}{x^2 + 3x} = \frac{(x - 5)(x + 3)}{x(x + 3)} = \frac{x - 5}{x}$$

15. What is $\dfrac{6}{4a} - \dfrac{2}{3a}$ expressed in simplest form?

(1) $\dfrac{4}{a}$ *(2) $\dfrac{5}{6a}$ (3) $\dfrac{8}{7a}$ (4) $\dfrac{10}{12a}$

$$\frac{6}{4a} - \frac{2}{3a} = \frac{6 \cdot 3}{4a \cdot 3} - \frac{2 \cdot 4}{3a \cdot 4} = \frac{18}{12a} - \frac{8}{12a} = \frac{18 - 8}{12a} = \frac{10}{12a} = \frac{5}{6a}$$

16. Which expression represents $\dfrac{-14a^2c^8}{7a^3c^2}$ in simplest form?

(1) $-2ac^4$ (2) $-2ac^6$ (3) $\dfrac{-2c^4}{a}$ *(4) $\dfrac{-2c^6}{a}$

$$\frac{-14a^2c^8}{7a^3c^2} = \frac{-14}{7} \bullet \frac{a^2}{a^3} \bullet \frac{c^8}{c^2} = \frac{-2c^6}{a}$$

17. What is the sum of $\dfrac{-x+7}{2x+4}$ and $\dfrac{2x+5}{2x+4}$?

*(1) $\dfrac{x+12}{2x+4}$ (2) $\dfrac{3x+12}{2x+4}$ (3) $\dfrac{x+12}{4x+8}$ (4) $\dfrac{3x+12}{4x+8}$

$$\frac{-x+7}{2x+4} + \frac{2x+5}{2x+4} = \frac{-x+7+2x+5}{2x+4} = \frac{x+12}{2x+4}$$

18. What is $\dfrac{\sqrt{32}}{4}$ expressed in simplest radical form?

*(1) $\sqrt{2}$ (2) $4\sqrt{2}$ (3) $\sqrt{8}$ (4) $\dfrac{\sqrt{8}}{2}$

$$\frac{\sqrt{32}}{4} = \frac{\sqrt{16\bullet 2}}{4} = \frac{4\sqrt{2}}{4} = \sqrt{2}$$

Remember these perfect squares: 4, 9, 16, 25, 36, 49, 64, 81, 100, $\cdots$

19. The expression $6\sqrt{50} + 6\sqrt{2}$ written in simplest radical form is

(1) $6\sqrt{52}$ (2) $13\sqrt{52}$ (3) $17\sqrt{2}$ *(4) $36\sqrt{2}$

$$6\sqrt{50} + 6\sqrt{2} = 6\sqrt{25\bullet 2} + 6\sqrt{2} = 6\bullet 5\sqrt{2} + 6\sqrt{2} = 30\sqrt{2} + 6\sqrt{2} = 36\sqrt{2}$$

20. The expression $\sqrt{72} - 3\sqrt{2}$ written in simplest radical form is

(1) $5\sqrt{2}$ (2) $3\sqrt{6}$ *(3) $3\sqrt{2}$ (4) $\sqrt{6}$

$$\sqrt{72} - 3\sqrt{2} = \sqrt{36\bullet 2} - 3\sqrt{2} = 6\sqrt{2} - 3\sqrt{2} = 3\sqrt{2}$$

II. Algebraic Expressions and Operations

21. What is half of 2^6?

(1) 1^3 (2) 1^6 (3) 2^3 *(4) 2^5

$$\frac{2^6}{2} = \frac{2^6}{2^1} = 2^{6-1} = 2^5$$

22. Which expression is equivalent to $(3x^2)^3$?

(1) $9x^5$ (2) $9x^6$ (3) $27x^5$ *(4) $27x^6$

$$(3x^2)^3 = 3^3 \bullet x^{2 \bullet 3} = 27x^6$$

23. Which expression is equivalent to $3^3 \bullet 3^4$?

(1) 9^{12} (2) 9^7 (3) 3^{12} *(4) 3^7

$$3^3 \bullet 3^4 = 3^{3+4} = 3^7$$

24. What is the product of 8.4×10^8 and 4.2×10^3 written in scientific notation?

(1) 2.0×10^5 (2) 12.6×10^{11} (3) 35.28×10^{11} *(4) 3.528×10^{12}

$(8.4 \times 10^8)(4.2 \times 10^3) = (8.4 \times 4.2)(10^8 \times 10^3) = 35.28 \times 10^{11} = 3.528 \times 10^{12}$

In scientific notation: $a \times 10^n$ where $1 \le a < 10$ and n is an integer.

25. What is the product of 12 and 4.2×10^6 expressed in scientific notation?

(1) 50.4×10^6 (2) 50.4×10^7 (3) 5.04×10^6 *(4) 5.04×10^7

$(12 \times 4.2) \times 10^6 = 50.4 \times 10^6 = 5.04 \times 10^7$

26. The length of a rectangular room is 7 less than three times the width, w, of the room. Which expression represents the area of the room?

(1) $3w - 4$ (2) $3w - 7$ (3) $3w^2 - 4w$ *(4) $3w^2 - 7w$

Length L = 3W - 7

Area A = L $\bullet$ W = (3W - 7) W = $3W^2 - 7W$

27. If $a + ar = b + r$, the value of a in terms of b and r can be expressed as

(1) $\dfrac{b}{r} + 1$ (2) $\dfrac{1+b}{r}$ *(3) $\dfrac{b+r}{1+r}$ (4) $\dfrac{1+b}{r+b}$

$a + ar = b + r$

$a(1 + r) = b + r$

$a = \dfrac{b + r}{1 + r}$

---

28. An example of an algebraic expression is

(1) $\dfrac{2x+3}{7} = \dfrac{13}{x}$ (3) $4x - 1 = 4$

*(2) $(2x+1)(x-7)$ (4) $x = 2$

The rest are equations.

---

29. Which verbal expression is represented by $\dfrac{1}{2}(n-3)$?

(1) one-half n decreased by 3   (3) the difference of one-half n and 3
(2) one-half n subtracted from 3   ***(4) one-half the difference of n and 3**

30. A formula used for calculating velocity is $v = \dfrac{1}{2}at^2$. What is a expressed in terms of v and t?

(1) $a = \dfrac{2v}{t}$ *(2) $a = \dfrac{2v}{t^2}$ (3) $a = \dfrac{v}{t}$ (4) $a = \dfrac{v}{2t^2}$

$v = \dfrac{1}{2}at^2$

$2v = at^2$

$a = \dfrac{2v}{t^2}$

II. Algebraic Expressions and Operations

Show Work:

1. Factor completely: $4x^3 - 36x$

$= 4x(x^2 - 9)$ Find the common factors first.
$= 4x(x + 3)(x - 3)$

---

2. Simplify: $\dfrac{27k^5 m^8}{(4k^3)(9m^2)}$

$= \dfrac{27}{4 \bullet 9} \bullet \dfrac{k^5}{k^3} \bullet \dfrac{m^8}{m^2} = \dfrac{3k^2 m^6}{4}$

---

3. Perform the indicated operation and simplify: $\dfrac{3x+6}{4x+12} \div \dfrac{x^2-4}{x+3}$

$= \dfrac{3x + 6}{4x + 12} \bullet \dfrac{x + 3}{x^2 - 4}$ multiply inverse

$= \dfrac{3(x + 2)}{4(x + 3)} \bullet \dfrac{(x + 3)}{(x + 2)(x - 2)}$ factor before simplifying

$= \dfrac{3}{4(x - 2)}$

---

4. Express in simplest form: $\dfrac{x^2 + 9x + 14}{x^2 - 49} \div \dfrac{3x+6}{x^2+x-56}$

$= \dfrac{x^2 + 9x + 14}{x^2 - 49} \bullet \dfrac{x^2 + x - 56}{3x + 6} = \dfrac{(x + 2)(x + 7)}{(x + 7)(x - 7)} \bullet \dfrac{(x + 8)(x - 7)}{3(x + 2)} = \dfrac{x + 8}{3}$

---

5. Express $5\sqrt{72}$ in simplest radical form.

$5\sqrt{72} = 5\sqrt{36 \bullet 2} = 5 \bullet 6\sqrt{2} = 30\sqrt{2}$

---

1. Pam is playing with red and black marbles. The number of red marbles she has is three more than twice the number of black marbles she has. She has 42 marbles in all. How many red marbles does Pam have?
(1)   13   (2)   15   *(3)   29   (4)   33

> B: the number of the black marbles; R: the number of the red marbles, $R = 2B + 3$
> $2B + 3 + B = 42$, $3B = 39$, $B = 13$, $\mathbf{R = 29}$

2. Sam and Odel have been selling frozen pizzas for a class fundraiser. Sam has sold half as many pizzas as Odel. Together they have sold a total of 126 pizzas. How many pizzas did Sam sell?
(1)   21   *(2)   42   (3)   63   (4)   84

> x: the number of pizzas that Sam sold;   2x: the number of pizzas that Odel sold
> $x + 2x = 126$, $3x = 126$, $\mathbf{x = 42}$

3. What is the speed, in meters per second, of a paper airplane that flies 24 meters in 6 seconds?
(1)   144   (2)   30   (3)   18   *(4)   4

> speed $\; s = \dfrac{d}{t} = \dfrac{24}{6} = \mathbf{4}$

4. Tamara has a cell phone plan that charges $0.07 per minute plus a monthly fee of $19.00. She budgets $29.50 per month for total cell phone expenses without taxes. What is the maximum number of minutes Tamara could use her phone each month in order to stay within her budget?
*(1)   150   (2)   271   (3)   421   (4)   692

> x: the number of minutes
> $19 + 0.07x \le 29.50$, $0.07x \le 10.5$, $\mathbf{x \le 150}$

5. Rhonda has $1.35 in nickels and dimes in her pocket. If she has six more dimes than nickels, which equation can be used to determine x, the number of nickels she has?
(1) $0.05(x+6)+0.10x = 1.35$ (3) $0.05 + 0.10(6x) = 1.35$
*(2) $0.05x + 0.10(x+6) = 1.35$ (4) $0.15(x+6) = 1.35$

> each nickel: 0.05; each dime: 0.10

6. If the speed of sound is 344 meters per second, what is the approximate speed of sound, in meters per hour?

> 60 seconds = 1 minute
> 60 minutes = 1 hour

(1)   20,640   (2)   41,280   (3)   123,840   *(4)   **1,238,400**

---

1 hour = 60•60 = 3600 seconds
344•3600 = 1,238,400

---

7. At Genesee High School, the sophomore class has 60 more students than the freshman class. The junior class has 50 fewer students than twice the students in the freshman class. The senior class is three times as large as the freshman class. If there are a total of 1,424 students at Genesee High School, how many students are in the freshman class?

*(1)   **202**   (2)   205   (3)   235   (4)   236

---

x: the number of freshmen
x + (x + 60) + (2x - 50) + 3x = 1424
7x + 10 = 1424,   7x = 1414,   **x = 202**

---

8. Steve ran a distance of 150 meters in $1\frac{1}{2}$ minutes. What is his speed in meters per hour?

(1)   6   (2)   60   (3)   100   *(4)   **6,000**

$$s = \frac{d}{t} = \frac{150}{1.5} = 100 \text{ m/min}$$

1 hour = 60 minutes
s = 100•60 = 6,000 m/h

---

9. An electronics store sells DVD players and cordless telephones. The store makes a $75 profit on the sale of each DVD player (*d*) and a $30 profit on the sale of each cordless telephone (*c*). The store wants to make a profit of at least $255.00 from its sales of DVD players and cordless phones. Which inequality describes this situation?

(1) $75d + 30c < 255$ (3) $75d + 30c > 255$
(2) $75d + 30c \leq 255$ *(4) $75d + 30c \geq 255$

10. Which value of *x* is in the solution set of the inequality $-2x + 5 > 17$?

*(1)   −8   (2)   −6   (3)   −4   (4)   12

-2x + 5 > 17
-2x > 12
x < -6   when the inequality is divided by a negative number, the inequality sign is reversed.

11. Students in a ninth grade class measured their heights, h, in centimeters. The height of the shortest student was 155 cm, and the height of the tallest student was 190 cm. Which inequality represents the range of heights?

(1) $155 < h < 190$

*(2) $155 \leq h \leq 190$

(3) $h \geq 155$ or $h \leq 190$

(4) $h > 155$ or $h < 190$

12. The sign shown below is posted in front of a roller coaster ride at the Wadsworth County Fairgrounds.

All riders **MUST** be
at least 48 inches tall.

If h represents the height of a rider in inches, what is a correct translation of the statement on this sign?

(1) $h < 48$ (2) $h > 48$ (3) $h \leq 48$ *(4) $h \geq 48$

13. Which value of x is in the solution set of $\frac{4}{3}x + 5 < 17$?

*(1)   **8**   (2)   9   (3)   12   (4)   16

$$\frac{4}{3}x < 12$$
$$x < 9$$

14. The length of a rectangular window is 5 feet more than its width, w. The area of the window is 36 square feet. Which equation could be used to find the dimensions of the window?

(1) $w^2 + 5w + 36 = 0$

(2) $w^2 - 5w - 36 = 0$

(3) $w^2 - 5w + 36 = 0$

*(4) $w^2 + 5w - 36 = 0$

$(w + 5) \bullet w = 36,   w^2 + 5w = 36,   w^2 + 5w - 36 = 0$

15. What are the roots of the equation $x^2 - 10x + 21 = 0$?

(1) 1 and 21

(2) -5 and -5

*(3) **3 and 7**

(4) -3 and -7

$(x - 3)(x - 7) = 0,   x = 3$ and $x = 7$

III. Equations and Inequalities

16. When 36 is subtracted from the square of a number, the result is five times the number. What is the positive solution?

*(1)   9   (2)   6   (3)   3   (4)   4

$x^2 - 36 = 5x$
$x^2 - 5x - 36 = 0$
$(x - 9)(x + 4) = 0$
$x = 9$, $x = -4$

17. Which value of x is a solution of $\dfrac{5}{x} = \dfrac{x+13}{6}$?

(1)   -2   (2)   -3   (3)   -10   ***(4)   -15**

$x(x + 13) = 5 \bullet 6$ cross-multipy
$x^2 + 13x - 30 = 0$
$(x + 15)(x - 2) = 0$
$x = -15$, $x = 2$

18. What is the value of x in the equation $\dfrac{2}{x} - 3 = \dfrac{26}{x}$?

*(1) -8 (2) $-\dfrac{1}{8}$ (3) $\dfrac{1}{8}$ (4)   8

$2 - 3x = 26$ multiply x on both sides
$-3x = 24$, $x = -8$

19. Which value of x is the solution of $\dfrac{x}{3} + \dfrac{x+1}{2} = x$?

(1)   1   (2)   -1   ***(3)   3**   (4)   -3

$2x + 3(x + 1) = 6x$ multiply LCD: 6 on both sides
$2x + 3x + 3 = 6x$, $x = 3$

20. The sum of two numbers is 47, and their difference is 15. What is the larger number?

(1)   16   ***(2)   31**   (3)   32   (4)   36

$x + y = 47$ (1)
$x - y = 15$ (2)
add Eq. (1) and Eq. (2)
$2x = 62$, $x = 31$, $y = 16$

21. Jack bought 3 slices of cheese pizza and 4 slices of mushroom pizza for a total cost of $12.50. Grace bought 3 slices of cheese pizza and 2 slices of mushroom pizza for a total cost of $8.50. What is the cost of one slice of mushroom pizza?

(1) $1.50   *(2) $2.00   (3) $3.00   (4) $3.50

C: the cost of one slice of cheese pizza
M: the cost of one slice of the mushroom pizza
3C + 4M = 12.50   (1)
3C + 2M = 8.50   (2)
subtratct Eq. (2) from Eq. (1)
2M = 4.00, **M = 2.00**

---

22. What is the value of the y-coordinate of the solution to the system of equations $x + 2y = 9$ and $x - y = 3$?

(1)   6   *(2)   2   (3)   3   (4)   5

---

x + 2y = 9   (1)
x - y = 3   (2)
subtratct Eq. (2) from Eq. (1)
3y = 6,   **y = 2**

---

23. Julia went to the movies and bought one jumbo popcorn and two chocolate chip cookies for $5.00. Marvin went to the same movie and bought one jumbo popcorn and four chocolate chip cookies for $6.00. How much does one chocolate chip cookie cost?

*(1)   **$0.50**   (2) $0.75   (3) $1.00   (4) $2.00

P: the cost of one popcon
C: the cost of one cookie
P + 2C = 5.00   (1)
P + 4C = 6.00   (2)
subtract Eq. (1) from Eq. (2)
2C = 1.00,   **C = 0.50**

---

24. Which ordered pair is a solution of the system of equations $y = x^2 - x - 20$ and $y = 3x - 15$?

(1) $(-5, -30)$ *(2) $(-1, -18)$ (3) $(0, 5)$ (4) $(5, -1)$

Plug in the numbers in both equations.
Test the easiest one first, like (0, 5), (-1, -18), •••

III. Equations and Inequalities

Show Work:

1. Hannah took a trip to visit her cousin. She drove 120 miles to reach her cousin's house and the same distance back home. It took her 1.2 hours to get halfway to her cousin's house. What was her average speed, in miles per hour, for the first 1.2 hours of the trip? Hannah's average speed for the remainder of the trip to her cousin's house was 40 miles per hour. How long, in hours, did it take her to drive the remaining distance? Traveling home along the same route, Hannah drove at an average rate of 55 miles per hour. After 2 hours her car broke down. How many miles was she from home?

(1). The halfway of 120 miles is 60 miles.

$$s = \frac{d}{t} = \frac{60}{1.2} = 50 \text{ mph (miles per hour)}$$

(2). $t = \dfrac{d}{s} = \dfrac{60}{40} = 1.5$ hours

(3). $d = s \bullet t = 55 \bullet 2 = 110$ miles

Distance from home: 120 - 110 = 10 miles

2. Peter begins his kindergarten year able to spell 10 words. He is going to learn to spell 2 new words every day. Write an inequality that can be used to determine how many days, d, it takes Peter to be able to spell *at least* 75 words. Use this inequality to determine the minimum number of whole days it will take for him to be able to spell *at least* 75 words.

d: the number of days

$10 + 2d \geq 75$

$2d \geq 65$

$d \geq 32.5$

minimum 33 days

3. Given: $A = \{18, 6, -3, -12\}$

Determine all elements of set A that are in the solution of the inequality $\frac{2}{3}x + 3 < -2x - 7$.

$\dfrac{2}{3}x + 2x < -10$

$\dfrac{8}{3}x < -10$

$x < -\dfrac{30}{8}$ that is $x < -3.75$

the solution set: $\{ -12 \}$

4. A contractor needs 54 square feet of brick to construct a rectangular walkway. The length of the walkway is 15 feet more than the width. Write an equation that could be used to determine the dimensions of the walkway. Solve this equation to find the length and width, in feet, of the walkway.

$L = W + 15$
$W(W + 15) = 54$
$W^2 + 15W - 54 = 0$
$(W + 18)(W - 3) = 0$
$W = 3$ (W = -18 rejected)
$L = 3 + 15 = 18$

5. Find three consecutive positive even integers such that the product of the second and third integers is twenty more than ten times the first integer. [Only an algebraic solution can receive full credit.]

Three consecutive positive even integers: $x, x + 2, x + 4$
$(x + 2)(x + 4) = 10x + 20$
$x^2 + 6x + 8 = 10x + 20$
$x^2 - 4x - 12 = 0$
$(x - 6)(x + 2) = 0$
$x = 6$ (x = -2   rejected)
$\{ 6, 8, 10 \}$

6. Solve for x: $\dfrac{x+1}{x} = \dfrac{-7}{x-12}$

$(x + 1)(x - 12) = -7x$ cross-multiply
$x^2 - 11x - 12 = -7x$
$x^2 - 4x - 12 = 0$
$(x - 6)(x + 2) = 0$
$x = 6, \quad x = -2$ must check: both are the solutions

7. Solve the following system of equations algebraically:

$$3x + 2y = 4 \qquad (1)$$
$$4x + 3y = 7 \qquad (2)$$

[Only an algebraic solution can receive full credit.]

$9x + 6y = 12$ (3)   Eq. (1) x 3
$8x + 6y = 14$ (4)   Eq. (2) x 2
$\mathbf{x = -2}$ subtratct Eq. (4) from Eq. (3)
$3(-2) + 2y = 4$ substitute x by -2 in Eq. (1)
$2y = 10, \quad \mathbf{y = 5}$
Solution: (-2, 5)

IV. Right Triangles and Trigonometry

1. What is the value of *x*, in inches, in the right triangle below?

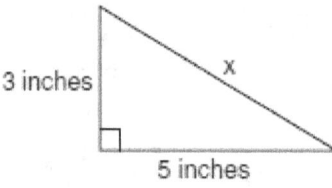

(1) $\sqrt{15}$ (2) 8 *(3) $\sqrt{34}$ (4) 4

$$x = \sqrt{3^2 + 5^2} = \sqrt{34}$$

2. The center pole of a tent is 8 feet long, and a side of the tent is 12 feet long as shown in the diagram below.

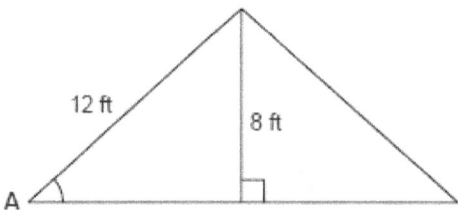

If a right angle is formed where the center pole meets the ground, what is the measure of angle *A* to the *nearest degree*?

(1) 34 *(2) 42 (3) 48 (4) 56

$$\sin A = \frac{8}{12}, \quad m\angle A = \sin^{-1}\left(\frac{8}{12}\right) \approx 42$$

3. In the right triangle shown in the diagram below, what is the value of *x* to the *nearest whole number*?

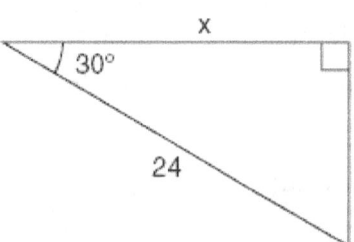

(1) 12 (2) 14 *(3) 21 (4) 28

$$\cos 30° = \frac{x}{24}, \quad x = 24 \bullet \cos 30° \approx 21$$

4. The diagram below shows right triangle *UPC*.

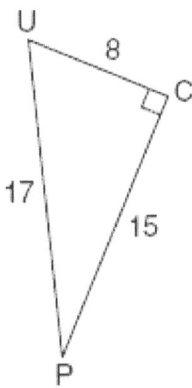

Which ratio represents the sine of ∠*U*?

(1) $\dfrac{15}{8}$ *(2) $\dfrac{15}{17}$ (3) $\dfrac{8}{15}$ (4) $\dfrac{8}{17}$

5. Which equation shows a correct trigonometric ratio for angle *A* in the right triangle below?

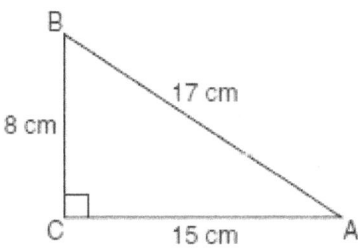

(1) $\sin A = \dfrac{15}{17}$ (2) $\tan A = \dfrac{8}{17}$ *(3) $\cos A = \dfrac{15}{17}$ (4) $\tan A = \dfrac{5}{8}$

6. In △*ABC*, the measure of ∠*B* = 90°, *AC* = 50, *AB* = 48, and *BC* = 14. Which ratio represents the tangent of ∠*A*?

(1) $\dfrac{14}{50}$ *(2) $\dfrac{14}{48}$ (3) $\dfrac{48}{50}$ (4) $\dfrac{48}{14}$

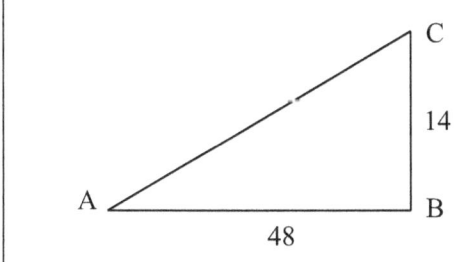

m∠B = 90, AC is the hypotenuse.

$$\tan A = \dfrac{\text{Opp}}{\text{Adj}} = \dfrac{14}{48}$$

Show Work:

1. A stake is to be driven into the ground away from the base of a 50-foot pole, as shown in the diagram below. A wire from the stake on the ground to the top of the pole is to be installed at an angle of elevation of 52°.

How far away from the base of the pole should the stake be driven in, to the *nearest foot*? What will be the length of the wire from the stake to the top of the pole, to the *nearest foot*?

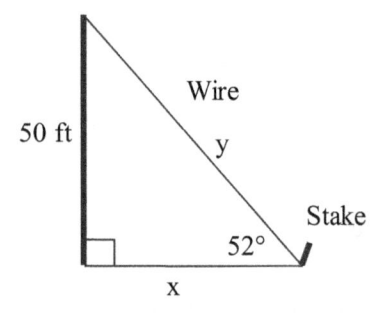

x: the distance between the pole and the stake
y: the length of the wire

(1). $\tan 52° = \dfrac{50}{x}$, $x = \dfrac{50}{\tan 52°} \approx 39$

(2). $\sin 52° = \dfrac{50}{y}$, $y = \dfrac{50}{\sin 52°} \approx 63$

2. In right triangle ABC, $AB = 20$, $AC = 12$, $BC = 16$, and $m\angle C = 90$. Find, to the *nearest degree*, the measure of $\angle A$.

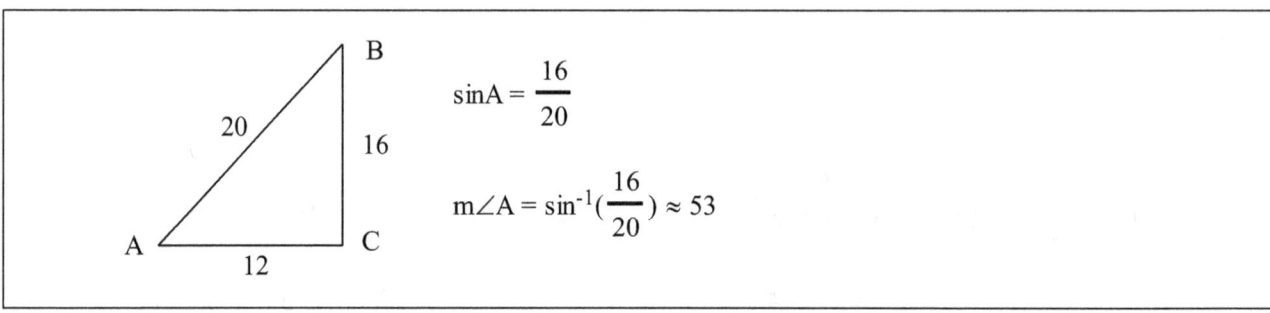

$\sin A = \dfrac{16}{20}$

$m\angle A = \sin^{-1}\left(\dfrac{16}{20}\right) \approx 53$

3. A communications company is building a 30-foot antenna to carry cell phone transmissions. As shown in the diagram below, a 50-foot wire from the top of the antenna to the ground is used to stabilize the antenna.

Find, to the *nearest degree,* the measure of the angle that the wire makes with the ground.

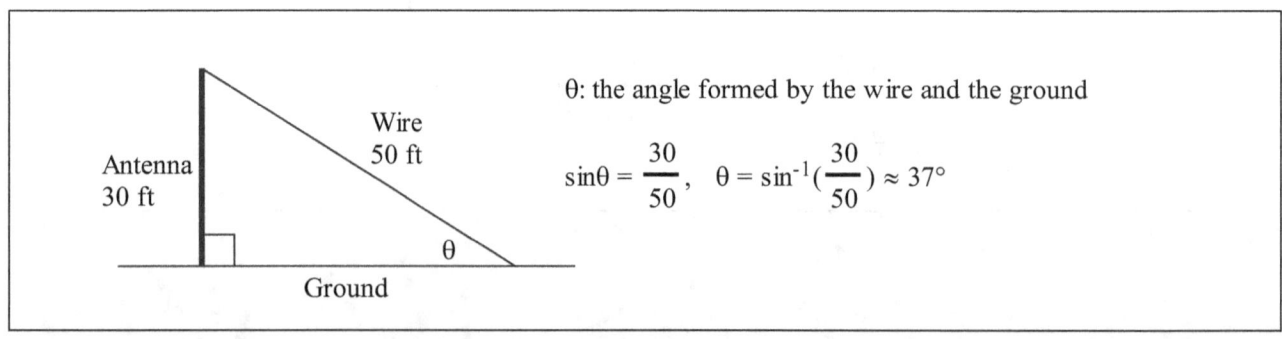

θ: the angle formed by the wire and the ground

$\sin \theta = \dfrac{30}{50}$, $\theta = \sin^{-1}\left(\dfrac{30}{50}\right) \approx 37°$

1. Luis is going to paint a basketball court on his driveway, as shown in the diagram below. This basketball court consists of a rectangle and a semicircle.

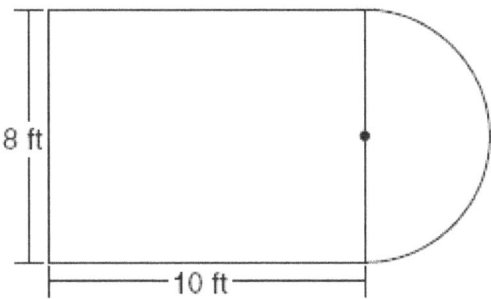

Which expression represents the area of this basketball court, in square feet?
(1) 80   (3) $80 + 16\pi$
*(2) **80 + 8π**   (4) $80 + 64\pi$

> The area of the rectangle = $8 \bullet 10 = 80$;
> The area of the semicircle = $\dfrac{1}{2}\pi r^2 = \dfrac{1}{2}\pi \bullet 4^2 = 8\pi$
> The total area is $80 + 8\pi$

2. Lenny made a cube in technology class. Each edge measured 1.5 cm. What is the volume of the cube in cubic centimeters?
(1)   2.25   *(2)   **3.375**   (3)   9.0   (4)   13.5

> $V = s^3 = 1.5^3 = 3.375$

3. How many square inches of wrapping paper are needed to entirely cover a box that is 2 inches by 3 inches by 4 inches?
(1)   18   (2)   24   (3)   26   *(4)   **52**

> Surface area of the box = $2(lw + hw + lh)$
> $= 2(2 \bullet 3 + 2 \bullet 4 + 3 \bullet 4)$
> $= 52$

4. A cylindrical container has a diameter of 12 inches and a height of 15 inches, as illustrated in the diagram below.

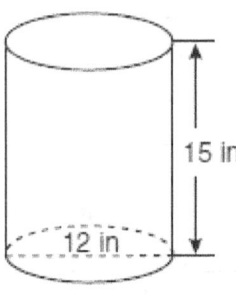

(Not drawn to scale)

What is the volume of this container to the *nearest tenth* of a cubic inch?
(1)   6,785.8   (2)   4,241.2   (3)   2,160.0   *(4)   **1,696.5**

$V = \pi r^2 h = \pi \bullet 6^2 \bullet 15 \approx 1696.5$

---

5. The groundskeeper is replacing the turf on a football field. His measurements of the field are 130 yards by 60 yards. The actual measurements are 120 yards by 54 yards. Which expression represents the relative error in the measurement?

*(1) $\dfrac{(130)(60) - (120)(54)}{(120)(54)}$ (3) $\dfrac{(130)(60) - (120)(54)}{(130)(60)}$

(2) $\dfrac{(120)(54)}{(130)(60) - (120)(54)}$ (4) $\dfrac{(130)(60)}{(130)(60) - (120)(54)}$

$$\text{relative error} = \frac{\text{absolute error}}{\text{actual value}}$$

---

6. To calculate the volume of a small wooden cube, Ezra measured an edge of the cube as 2 cm. The actual length of the edge of Ezra's cube is 2.1 cm. What is the relative error in his volume calculation to the *nearest hundredth*?
(1)   0.13   *(2)   **0.14**   (3)   0.15   (4)   0.16

The measured value of volume $= 2^3 = 8$
The actual value of volume $= 2.1^3 = 9.261$
$$\text{relative error} = \frac{|8 - 9.261|}{9.261} \approx \mathbf{0.14}$$

Show Work:

1. Serena's garden is a rectangle joined with a semicircle, as shown in the diagram below. Line segment AB is the diameter of semicircle P. Serena wants to put a fence around her garden. Calculate the length of fence Serena needs to the *nearest tenth of a foot*.

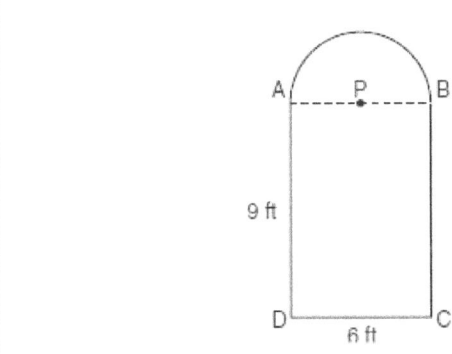

The length of the semicircle $= \pi r = \pi \bullet 3 = 3\pi$
The length of the fence $= 9 + 6 + 9 + 3\pi \approx$ **33.4**

2. A designer created the logo shown below. The logo consists of a square and four quarter-circles of equal size. Express, in terms of π, the exact area, in square inches, of the shaded region.

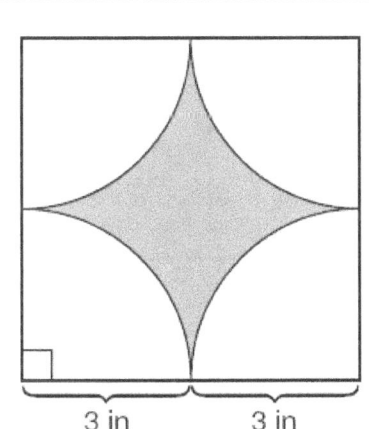

The area of four quarter-circles is equal to the area of a circle: $\pi \bullet 3^2 = 9\pi$
The area of the square: 6^2
The area of the shaded region $= 6^2 - 9\pi =$ **36 - 9π**

3. In the diagram below, the circumference of circle O is 16π inches. The length of $\overline{BC}$ is three-quarters of the length of diameter $\overline{AD}$ and $CE = 4$ inches. Calculate the area, in square inches, of trapezoid $ABCD$.

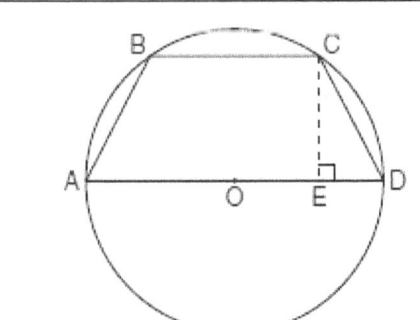

Circumference $C = \pi d = 16\pi$, $d = 16$
Diameter $AD = 16$
$BC = \dfrac{3}{4} AD = \dfrac{3 \bullet 16}{4} = 12$
Area of the trapezoid $A = \dfrac{b_1 + b_2}{2} \bullet h = \dfrac{12 + 16}{2} \bullet 4 =$ **56**

VI. Geometric Measurements

4. A soup can is in the shape of a cylinder. The can has a volume of 342 cm^3 and a diameter of 6 cm. Express the height of the can in terms of π. Determine the maximum number of soup cans that can be stacked on their base between two shelves if the distance between the shelves is exactly 36 cm. Explain your answer.

---

(1) $V = \pi r^2 \cdot h$

$342 = \pi \cdot 3^2 \cdot h$

$h = \dfrac{342}{9\pi} = \dfrac{\mathbf{38}}{\boldsymbol{\pi}}$

(2) $h = \dfrac{\mathbf{38}}{\pi} \approx 12.1$

The maximum number of soup cans that can

be stacked between two shelves:

$12.1 \cdot n \leq 36$

$n \leq 2.97$, $\mathbf{n = 2}$ (rounded to whole number)

---

5. Sophie measured a piece of paper to be 21.7 cm by 28.5 cm. The piece of paper is actually 21.6 cm by 28.4 cm. Determine the number of square centimeters in the area of the piece of paper using Sophie's measurements. Determine the number of square centimeters in the actual area of the piece of paper. Determine the relative error in calculating the area. Express your answer as a decimal to the *nearest thousandth*. Sophie does not think there is a significant amount of error. Do you agree or disagree? Justify your answer.

The measured value of area = 21.7 x 28.5 = 618.45
The actual value of area = 21.6 x 28.4 = 613.44

relative eror = $\dfrac{|21.7 \times 28.5 - 21.6 \times 28.4|}{21.6 \times 28.4} \approx 0.008$

The relatve error is very small, therefore Sophie is right about it.

---

6. Using his ruler, Howell measured the sides of a rectangular prism to be 5 cm by 8 cm by 4 cm. The actual measurements are 5.3 cm by 8.2 cm by 4.1 cm. Find Howell's relative error in calculating the volume of the prism, to the *nearest thousandth*.

The measured valued of volume V_m = 5 x 8 x 4
The actual value of volume V_a = 5.3 x 8.2 x 4.1

relative error = $\dfrac{|V_m - V_a|}{V_a} = \dfrac{|5 \times 8 \times 4 - 5.3 \times 8.2 \times 4.1|}{5.3 \times 8.2 \times 4.1} \approx 0.102$

---

7. Alexis calculates the surface area of a gift box as 600 square inches. The actual surface area of the gift box is 592 square inches. Find the relative error of Alexis' calculation expressed as a decimal to the *nearest thousandth*.

---

relative error = $\dfrac{|600 - 592|}{592} \approx 0.014$

1. What is the slope of the line containing the points $(3, 4)$ and $(-6, 10)$?

(1) $\dfrac{1}{2}$ (2)   2   ***(3)** $-\dfrac{2}{3}$ (4) $-\dfrac{3}{2}$

$$m = \frac{y_2 - y_1}{x_2 - x_1} = \frac{10 - 4}{-6 - 3} = \frac{6}{-9} = -\frac{2}{3}$$

2. What is an equation of the line that passes through the points $(3, -3)$ and $(-3, -3)$?

(1) $y = 3$ ***(3)** $y = -3$
(2) $x = -3$ (4) $x = y$

$y_1 = y_2 = -3$ It is a horizontal line: $y = -3$

3. Which equation represents a line that is parallel to the line $y = 3 - 2x$?

*(1) $4x + 2y = 5$ (3) $y = 3 - 4x$
(2) $2x + 4y = 1$ (4) $y = 4x - 2$

Rewrite the given line $y = 3 - 2x$ in the slope-intercept form: $y = -2x + 3$, slope: -2

Rewrite Eq. (1) in slope-intercept form: $y = -2x + \dfrac{5}{2}$, slope: -2

Parallel lines have the same slope.

4. What is an equation of the line that passes through the point $(4, -6)$ and has a slope of -3?

*(1) $y = -3x + 6$ (3) $y = -3x + 10$
(2) $y = -3x - 6$ (4) $y = -3x + 14$

All equations have a slope of -3. Test the point $(4, -6)$ in each equation.

5. Which equation represents the line that passes through the points $(-3, 7)$ and $(3, 3)$?

(1) $y = \dfrac{2}{3}x + 1$ ***(3)** $y = -\dfrac{2}{3}x + 5$

(2) $y = \dfrac{2}{3}x + 9$ (4) $y = -\dfrac{2}{3}x + 9$

$$m = \frac{7 - 3}{-3 - 3} = \frac{4}{-6} = -\frac{2}{3} ,$$ Test the point $(3, 3)$ in Eq. (3) and Eq. (4).

6. Which point is on the line $4y - 2x = 0$?

*(1) $(-2, -1)$ (2) $(-2, 1)$ (3) $(-1, -2)$ (4) $(1, 2)$

Test each point in the equation.

---

7. Which equation represents a line parallel to the graph of $2x - 4y = 16$?

*(1) $y = \frac{1}{2} x - 5$ (3) $y = -2x + 6$

(2) $y = -\frac{1}{2} x + 4$ (4) $y = 2x + 8$

Rewrite the equation 2x - 4y = 16 in the slope-intercept form: y = $\frac{1}{2}$x - 4, slope: $\frac{1}{2}$

---

8. Which linear equation represents a line containing the point $(1, 3)$?

(1) $x + 2y = 5$ *(3) $2x + y = 5$

(2) $x - 2y = 5$ (4) $2x - y = 5$

Test the point (1, 3) in each equation.

---

9. The graphs of the equations $y = 2x - 7$ and $y - kx = 7$ are parallel when k equals

(1)   -2   *(2)   2   (3)   -7   (4)   7

Rewrite the equation y - kx = 7 in the slope-intercept form: y = kx + 7 . Compare the slope: **k = 2**

---

10. It takes Tammy 45 minutes to ride her bike 5 miles. At this rate, how long will it take her to ride 8 miles?

(1)   0.89 hour   (2)   1.125 hours   (3)   48 minutes   *(4)   **72 minutes**

Direct variation: $\dfrac{x_1}{x_2} = \dfrac{y_1}{y_2}$, $\dfrac{45}{x} = \dfrac{5}{8}$,   5x = 360 ,   **x = 72 minutes**

---

11. Consider the graph of the equation $y = ax^2 + bx + c$, when $a \neq 0$. If a is multiplied by 3, what is true of the graph of the resulting parabola?

   (1) The vertex is 3 units above the vertex of the original parabola.

   (2) The new parabola is 3 units to the right of the original parabola.

   (3) The new parabola is wider than the original parabola.

** *(4) The new parabola is narrower than the original parabola.**

12. What are the vertex and axis of symmetry of the parabola $y = x^2 - 16x + 63$?

*(1) **vertex:** $(8, -1)$; **axis of symmetry:** $x = 8$ (3) vertex: $(-8, -1)$; axis of symmetry: $x = -8$

(2) vertex: $(8, 1)$; axis of symmetry: $x = 8$ (4) vertex: $(-8, 1)$; axis of symmetry: $x = -8$

The axis of symmetry: $x = \dfrac{-b}{2a} = \dfrac{16}{2} = 8$,   **x = 8**

Vertex: $x = 8$, $y = 8^2 - 16(8) + 63 = -1$,   **(8, -1)**

---

13. The equation $y = x^2 + 3x - 18$ is graphed on the set of axes below.

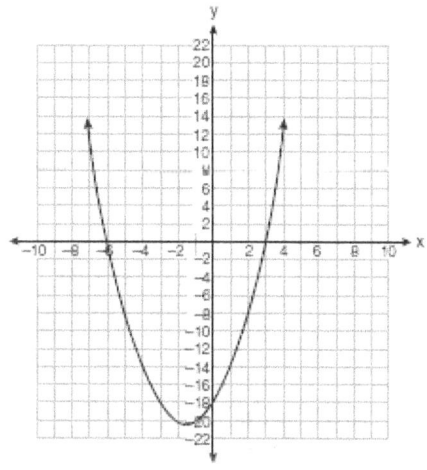

Based on this graph, what are the roots of the equation $x^2 + 3x - 18 = 0$?

(1) -3 and 6 *(3) **3 and -6**

(2) 0 and -18 (4) 3 and -18

The roots of the equation are x values of the x-intercepts (3, 0) and (-6, 0).

---

14. What is the equation of the axis of symmetry of the parabola shown in the diagram below?

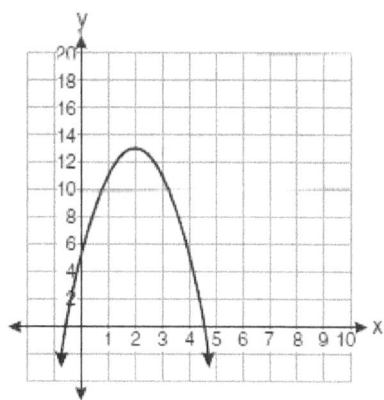

(1) $x = -0.5$ *(2) $x = 2$ (3) $x = 4.5$ (4) $x = 13$

15. What are the vertex and axis of symmetry of the parabola shown in the diagram below?

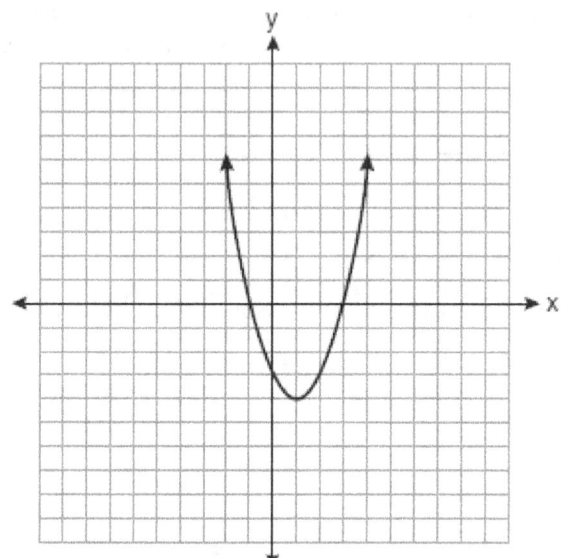

*(1) vertex: $(1, -4)$**; axis of symmetry:** $x = 1$ (3) vertex: $(-4, 1)$; axis of symmetry: $x = 1$
  (2) vertex: $(1, -4)$; axis of symmetry: $x = -4$ (4) vertex: $(-4, 1)$; axis of symmetry: $x = -4$

16. Daniel's Print Shop purchased a new printer for \$35,000. Each year it depreciates (loses value) at a rate of 5%. What will its approximate value be at the end of the fourth year?
(1) \$33,250.00   ***(3) \$28,507.72**
(2) \$30,008.13   (4) \$27,082.33

$$A = A_0(1 - 5\%)^4 = 35{,}000(0.95)^4 = 28507.72$$

17. Kathy plans to purchase a car that depreciates (loses value) at a rate of 14% per year. The initial cost of the car is \$21,000. Which equation represents the value, v, of the car after 3 years?
(1) $v = 21{,}000(0.14)^3$ (3) $v = 21{,}000(1.14)^3$
*(2) $v = 21{,}000(0.86)^3$ (4) $v = 21{,}000(0.86)(3)$

$$v = v_0(1 - 14\%)^3 = 21{,}000(0.86)^3$$

18. The value, y, of a \$15,000 investment over x years is represented by the equation $y = 15000(1.2)^{\frac{x}{3}}$. What is the profit (interest) on a 6-year investment?
*(1) \$6,600   (3) \$21,600
(2)   \$10,799   (4) \$25,799

$$y = 15{,}000(1.2)^{\frac{6}{3}} = 15{,}000(1.2)^2 = 21{,}600 ; \quad \text{Profit} = 21{,}600 - 15{,}000 = 6{,}600$$

19. In a science fiction novel, the main character found a mysterious rock that decreased in size each day. The table below shows the part of the rock that remained at noon on successive days.

Day	Fractional Part of the Rock Remaining
1	1
2	$\frac{1}{2}$
3	$\frac{1}{4}$
4	$\frac{1}{8}$

Which fractional part of the rock will remain at noon on day 7?

(1) $\frac{1}{128}$ *(2) $\frac{1}{64}$ (3) $\frac{1}{14}$ (4) $\frac{1}{12}$

---

day 1: $(\frac{1}{2})^0$,   day 2: $(\frac{1}{2})^1$,   day 3: $(\frac{1}{2})^2$,   day 4: $(\frac{1}{2})^3$, $\bullet\bullet\bullet$, day 7: $(\frac{1}{2})^6 = \frac{1}{64}$

---

20. Which ordered pair is a solution of the system of equations shown in the graph below?

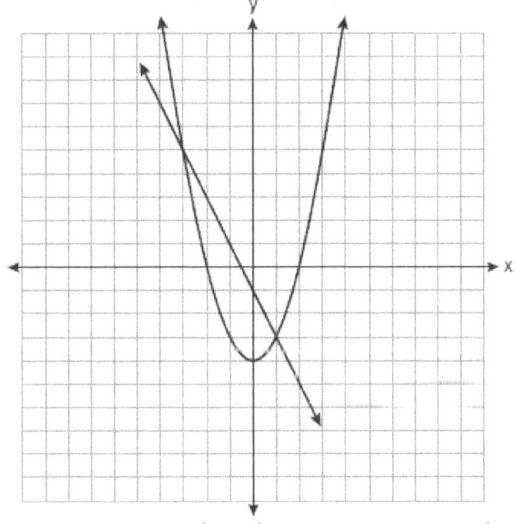

(1) $(-3, 1)$ *(2) $(-3, 5)$ (3) $(0, -1)$ (4) $(0, -4)$

The solution of a system of equations are the points of intersection (-3, 5) and (1, -3).

21. Which graph represents the solution of $3y - 9 \leq 6x$?

*(1)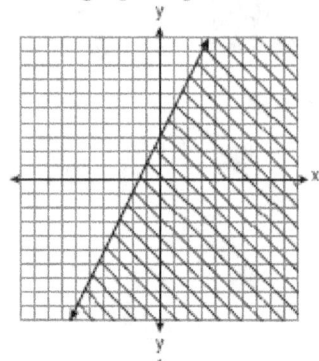

(3)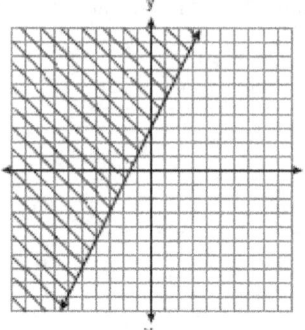

(2)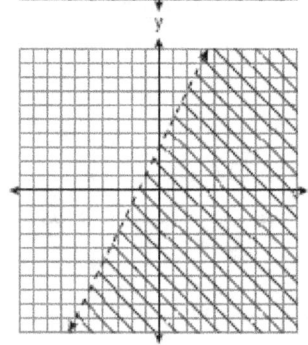

(4)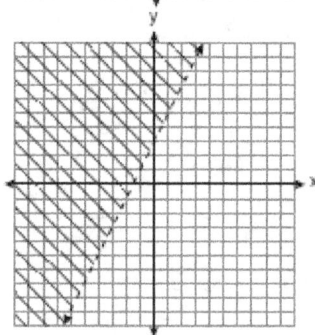

> Rewrite the inequality: y ≤ 2x + 3
> The solution is the region under y = 2x + 3 including the line (a solid line).

22. Which quadrant will be completely shaded in the graph of the inequality $y \leq 2x$?

(1) Quadrant I   (3) Quadrant III
(2) Quadrant II   *(4) **Quadrant IV**

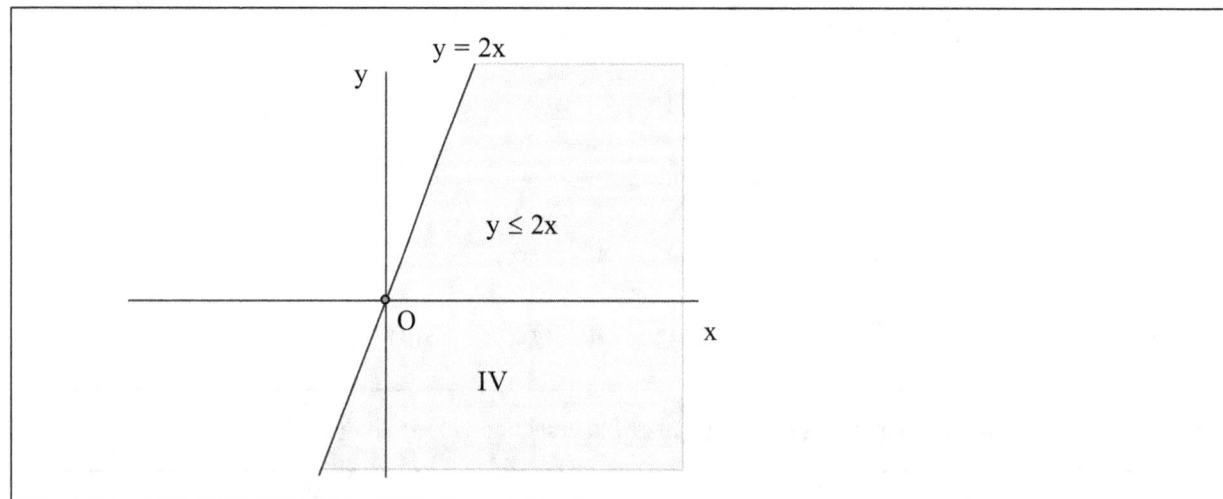

23. Which ordered pair is in the solution set of the system of linear inequalities graphed below?

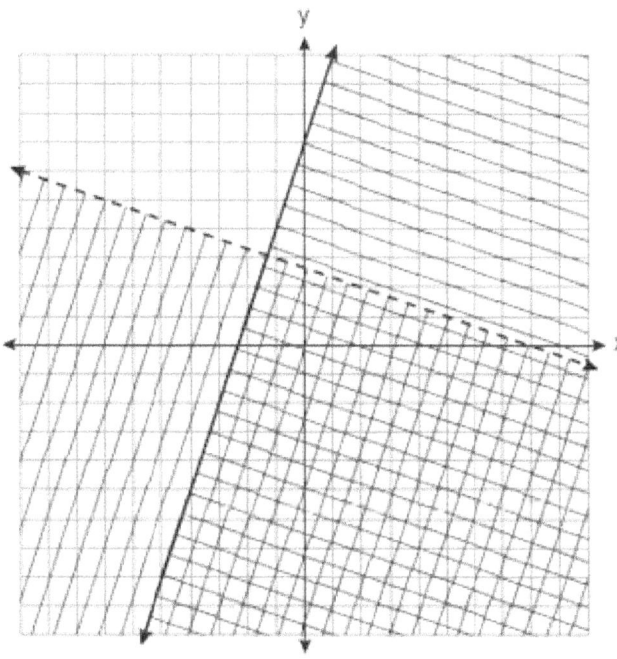

*(1)   (1, -4)   (2)   (-5, 7)   (3)   (5, 3)   (4)   (-7, -2)

24. Which ordered pair is in the solution set of the following system of linear inequalities?

$$y < 2x + 2$$
$$y \geq -x - 1$$

(1)   (0, 3)   *(2)   (2, 0)   (3)   (-1, 0)   (4)   (-1, -4)

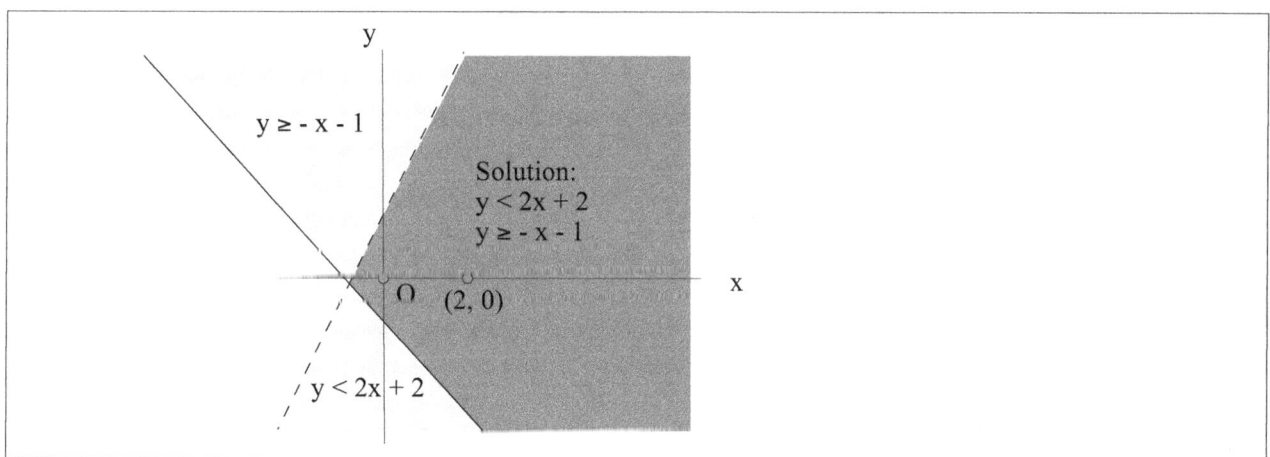

25. Which relation represents a function?

(1)   {(0, 3), (2, 4), (0, 6)}   (3)   {(2, 0), (6, 2), (6, -2)}
(2)   {(-7, 5), (-7, 1), (-10, 3), (-4, 3)}   *(4)   {(-6, 5), (-3, 2), (1, 2), (6, 5)}

The first elements in the ordered pairs can not repeat in a function.

26. Antwaan leaves a cup of hot chocolate on the counter in his kitchen. Which graph is the best representation of the change in temperature of his hot chocolate over time?

*(1)
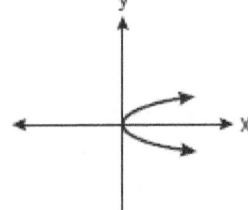

(3)
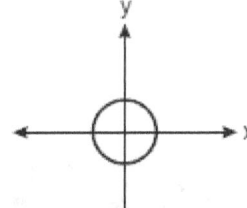

(2)
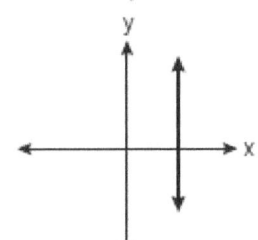

(4)
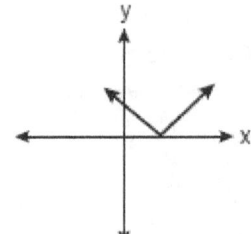

> The temperature of the hot chocolate reduces over time, but not in linear relation.

27. Which graph represents a function?

(1)
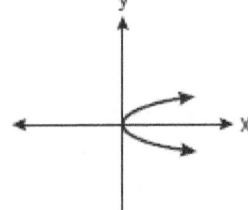

(3)
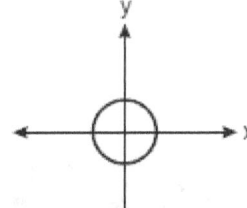

(2)
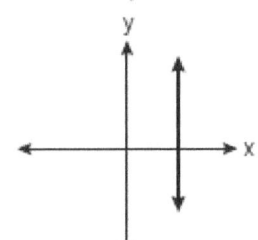

*(4)
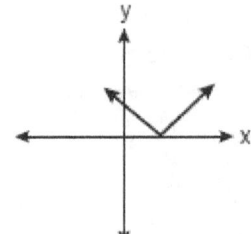

> Use the vertical line test: If any vertical line intersects the graph at one and only one point on its domain, then the relation is a function.

Show Work:

1. Graph and label the following equations on the set of axes below.

$$y = |x|$$

$$y = \left| \frac{1}{2} x \right|$$

Explain how *decreasing* the coefficient of x affects the graph of the equation $y = |x|$.

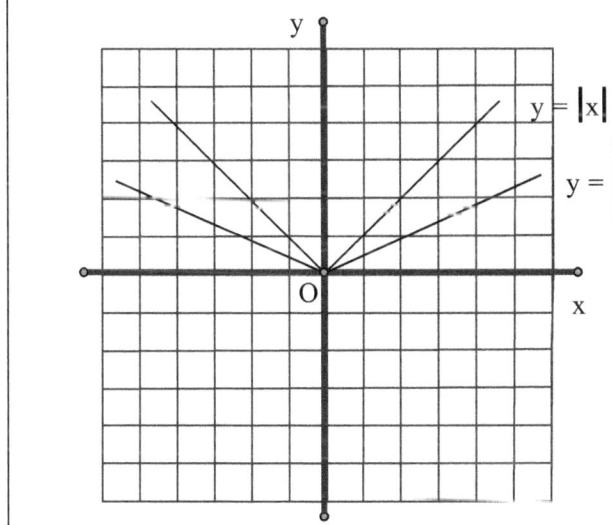

The graph of the equation $y = |x|$ becomes wider when the coefficient of x decreases.

2. Graph the equation $y = x^2 - 2x - 3$ on the accompanying set of axes. Using the graph, determine the roots of the equation $x^2 - 2x - 3 = 0$.

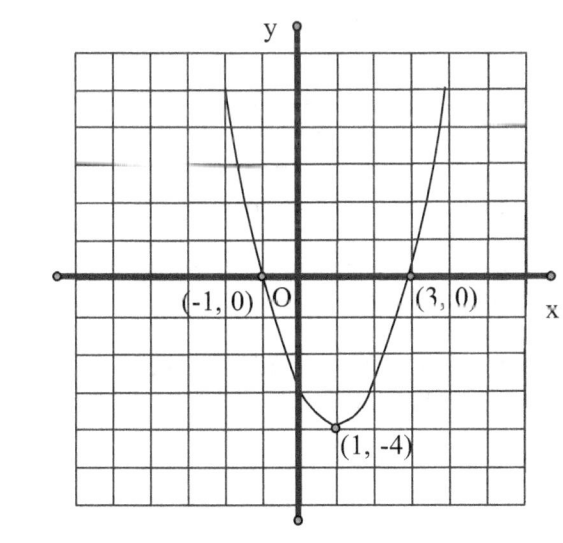

(1) Axis of symmetry:

$$x = \frac{-b}{2a} = \frac{2}{2 \bullet 1} = 1$$

(2) Vertex:

$$x = 1, \quad y = (1)^2 - 2(1) - 3 = -4$$
$$(1, -4)$$

(3) Opening:

a > 0, upward

(4) Properly choose a few points around the vertex.

x	-2	-1	0	1	2	3	4
y	5	0	-3	-4	-3	0	5

(5) The roots of the equation are the x values of the x-intercepts: -1 and 3

3. A bank is advertising that new customers can open a savings account with a $3\frac{3}{4}\%$ interest rate compounded annually. Robert invests \$5,000 in an account at this rate. If he makes no additional deposits or withdrawals on his account, find the amount of money he will have, to the *nearest cent*, after three years.

A_0: original amount \$5,000

A: amount after three years

$A = A_0(1 + 3\frac{3}{4}\%)^3 = 5,000 \bullet (1.0375)^3 = 5,583.86$

The amount will be \$5,583.86 after three years.

---

4. Solve the following systems of equations graphically, on the set of axes below, and state the coordinates of the point(s) in the solution set.

$$y = x^2 - 6x + 5$$

$$2x + y = 5$$

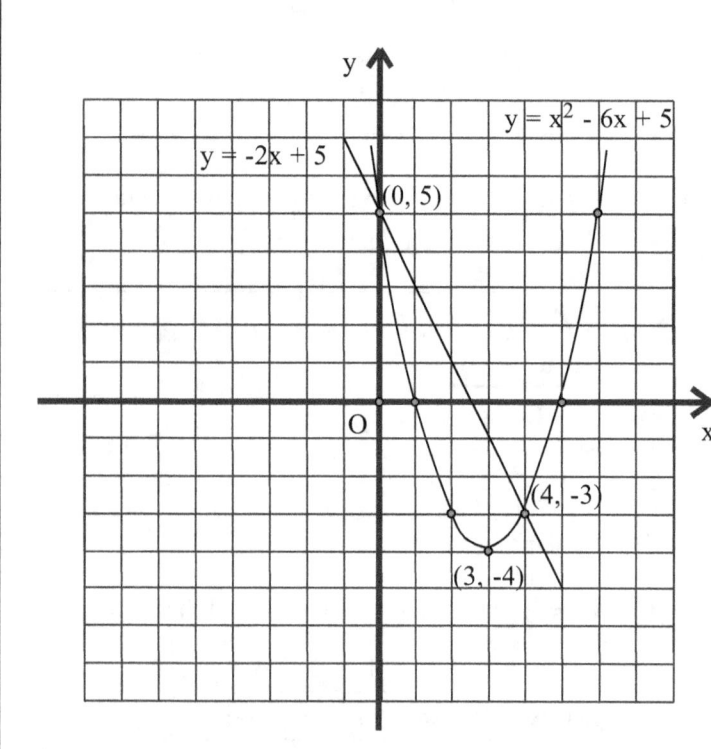

a. Graph the quadratic equation

(1) Find the vertex:

$$x = \frac{-b}{2a} = \frac{6}{2 \bullet 1} = 3$$

$$y = (3)^2 - 6(3) + 5 = -4$$

vertex : (3, -4)

(2) Choose a few points around the vertex:

x	0	1	2	3	4	5	6
y	5	0	-3	-4	-3	0	5

b. Graph the linear equation

Rewrite the equation in the slope-intercept form: y = -2x + 5

c. The points of intersection are the solution set: {(0, 5), (4, -3)}

5. Graph the solution set for the inequality $4x - 3y > 9$ on the set of axes below. Determine if the point $(1, -3)$ is in the solution set. Justify your answer.

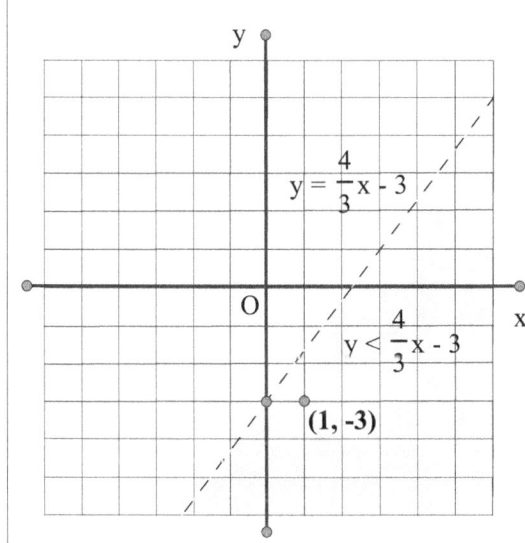

Rewrite the inequality:

$$-3y > -4x + 9 \quad \text{when divided by a negative number,}$$

$$y < \frac{4}{3}x - 3 \quad \text{the inequality sign is reversed.}$$

The solution of the $y < \frac{4}{3}x - 3$ is the region under

the line $y = \frac{4}{3}x - 3$.

Point $(1, -3)$ is located inside the region, therefore it is in the solution set.

6. On the set of axes below, graph the following system of inequalities and state the coordinates of a point in the solution set.

$$2x - y \geq 6$$
$$x > 2$$

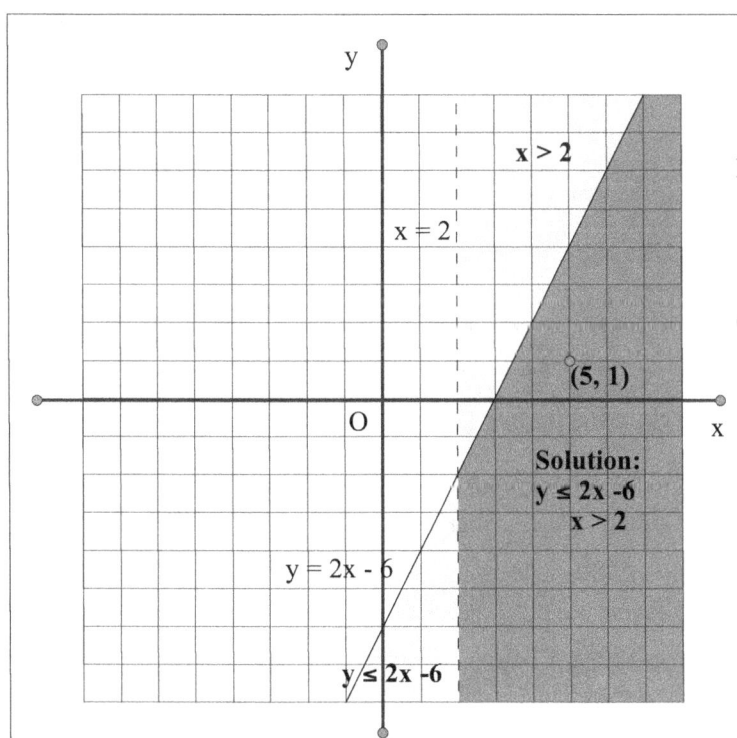

Rewrite the inequality:

$$2x - y \geq 6$$
$$-y \geq -2x + 6$$
$$y \leq 2x - 6 \quad \text{inequality sign reversed}$$

e.g. point $(5, 1)$ is a point in the solution set.

1. A spinner is divided into eight equal regions as shown in the diagram below.

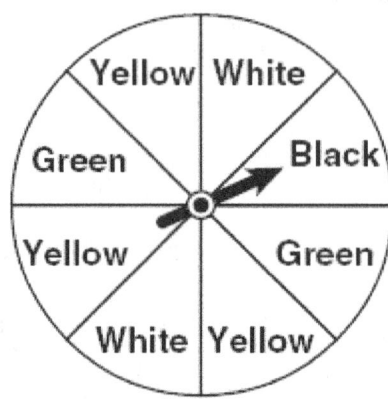

Which event is most likely to occur in one spin?
  (1) The arrow will land in a green or white area.
  (2) The arrow will land in a green or black area.
  (3) The arrow will land in a yellow or black area.
*(4) **The arrow will land in a yellow or green area.**

---

   3 Yellows, 2 Greens, 2 Whites, 1 Black
  (1) $\dfrac{2}{8} + \dfrac{2}{8} = \dfrac{4}{8}$ (2) $\dfrac{2}{8} + \dfrac{1}{8} = \dfrac{3}{8}$ (3) $\dfrac{3}{8} + \dfrac{1}{8} = \dfrac{4}{8}$ **(4) $\dfrac{3}{8} + \dfrac{2}{8} = \dfrac{5}{8}$**

---

2. The faces of a cube are numbered from 1 to 6. If the cube is tossed once, what is the probability that a prime number or a number divisible by 2 is obtained?

(1) $\dfrac{6}{6}$ *(2) $\dfrac{5}{6}$ (3) $\dfrac{4}{6}$ (4) $\dfrac{1}{6}$

---

   prime numbers: 2, 3, 5 ;   a number divisible by 2: 2, 4, 6
   Sample space: {1, 2, 3, 4, 5, 6} ;   Event: {2, 3, 4, 5, 6}
   $$P(E) = \dfrac{n(E)}{n(S)} = \dfrac{5}{6}$$

---

3. The faces of a cube are numbered from 1 to 6. If the cube is rolled once, which outcome is *least* likely to occur?
(1) rolling an odd number   (3) rolling a number less than 6
(2) rolling an even number   *(4) **rolling a number greater than 4**

---

   (1) {1, 3, 5}   (2) {2, 4, 6}   (3) {1, 2, 3, 4, 5}   **(4) {5, 6}**

4. Students in Ms. Nazzeer's mathematics class tossed a six-sided number cube whose faces are numbered 1 to 6. The results are recorded in the table below.

Result	Frequency
1	3
2	6
3	4
4	6
5	4
6	7

Based on these data, what is the empirical probability of tossing a 4?

(1) $\dfrac{8}{30}$ *(2) $\dfrac{6}{30}$ (3) $\dfrac{5}{30}$ (4) $\dfrac{1}{30}$

$$n(S) = 3 + 6 + 4 + 6 + 4 + 7 = 30; \quad n(4) = 6$$
$$P(4) = \dfrac{6}{30}$$

5. A bag contains eight green marbles, five white marbles, and two red marbles. What is the probability of drawing a red marble from the bag?

(1) $\dfrac{1}{15}$ *(2) $\dfrac{2}{15}$ (3) $\dfrac{2}{13}$ (4) $\dfrac{13}{15}$

6. Three high school juniors, Reese, Matthew, and Chris, are running for student council president. A survey is taken a week before the election asking 40 students which candidate they will vote for in the election. The results are shown in the table below.

Candidate's Name	Number of Students Supporting Candidate
Reese	15
Matthew	13
Chris	12

Based on the table, what is the probability that a student will vote for Reese?

(1) $\dfrac{1}{3}$ (2) $\dfrac{3}{5}$ *(3) $\dfrac{3}{8}$ (4) $\dfrac{5}{8}$

$$\dfrac{15}{40} = \dfrac{3}{8}$$

7. Keisha is playing a game using a wheel divided into eight equal sectors, as shown in the diagram below. Each time the spinner lands on orange, she will win a prize.

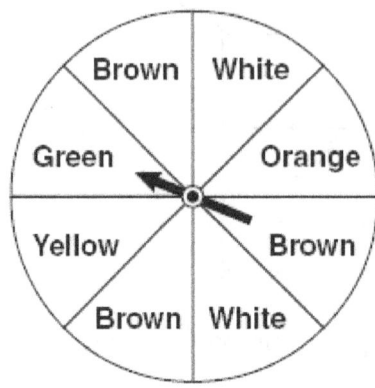

If Keisha spins this wheel twice, what is the probability she will win a prize on *both* spins?

*(1) $\frac{1}{64}$ (2) $\frac{1}{56}$ (3) $\frac{1}{16}$ (4) $\frac{1}{4}$

Use Counting Principle for probability: $\frac{1}{8} \cdot \frac{1}{8} = \frac{1}{64}$

---

8. How many different sandwiches consisting of one type of cheese, one condiment, and one bread choice can be prepared from five types of cheese, two condiments, and three bread choices?
(1)   10   (2)   13   (3)   15   *(4)   **30**

Use Counting Principle: $5 \cdot 2 \cdot 3 = 30$

---

9. John is going to line up his four golf trophies on a shelf in his bedroom. How many different possible arrangements can he make?
*(1)   **24**   (2)   16   (3)   10   (4)   4

$4! = 4 \cdot 3 \cdot 2 \cdot 1 = 24$

---

10. The bowling team at Lincoln High School must choose a president, vice president, and secretary. If the team has 10 members, which expression could be used to determine the number of ways the officers could be chosen?
(1) $_3P_{10}$ (2) $_7P_3$ *(3) $_{10}P_3$ (4) $_{10}P_7$

11. How many different three-letter arrangements can be formed using the letters in the word *ABSOLUTE* if each letter is used only once?
(1)   56   (2)   112   (3)   168   *(4)   **336**

$_8P_3 = 8 \cdot 7 \cdot 6 = 336$

Show Work:

1. Some books are laid on a desk. Two are English, three are mathematics, one is French, and four are social studies. Theresa selects an English book and Isabelle then selects a social studies book. Both girls take their selections to the library to read. If Truman then selects a book at random, what is the probability that he selects an English book?

Books left: 1 English, 3 Maths, 1 French, 3 Social Studies

$$P(E) = \frac{n(E)}{n(S)} = \frac{1}{8}$$

---

2. Jon is buying tickets for himself for two concerts. For the jazz concert, 4 tickets are available in the front row, and 32 tickets are available in the other rows. For the orchestra concert, 3 tickets are available in the front row, and 23 tickets are available in the other rows. Jon is randomly assigned one ticket for each concert. Determine the concert for which he is more likely to get a front-row ticket. Justify your answer.

For the jazz concert: $P(J) = \dfrac{4}{4 + 32} = \dfrac{4}{36} \approx 0.111$

For the orchestra concert: $P(O) = \dfrac{3}{3 + 23} = \dfrac{3}{26} \approx 0.115$

$P(O) > P(J)$

Jon is more likely to get a front-row ticket for the orchestra concert.

---

3. A password consists of three digits, 0 through 9, followed by three letters from an alphabet having 26 letters. **(1)** If repetition of digits is allowed, but repetition of letters is not allowed, determine the number of different passwords that can be made. **(2)** If repetition is not allowed for digits or letters, determine how many fewer different passwords can be made.

---

(1) Repetition of digits is allowed, but repetition of letters is not allowed.

$10 \bullet 10 \bullet 10 \bullet {}_{26}P_3 = 1000 \bullet 26 \bullet 25 \bullet 24 = 15,600,000$

(2) Repetition is not allowed for digits or letters.

${}_{10}P_3 \bullet {}_{26}P_3 = 10 \bullet 9 \bullet 8 \bullet 26 \bullet 25 \bullet 24 = 11,232,000$

The number of fewer passwords can be made:

$15,600,000 - 11,232,000 = 4,368,000$

4. A restaurant sells kids' meals consisting of one main course, one side dish, and one drink, as shown in the table below.

Kids' Meal Choices

Main Course	Side Dish	Drink
hamburger	French fries	milk
chicken nuggets	applesauce	juice
turkey sandwich		soda

Draw a tree diagram or list the sample space showing all possible kids' meals. **(1)** How many different kids' meals can a person order? **(2)** Jose does not drink juice. Determine the number of different kids' meals that do *not* include juice. **(3)** Jose's sister will eat *only* chicken nuggets for her main course. Determine the number of different kids' meals that include chicken nuggets.

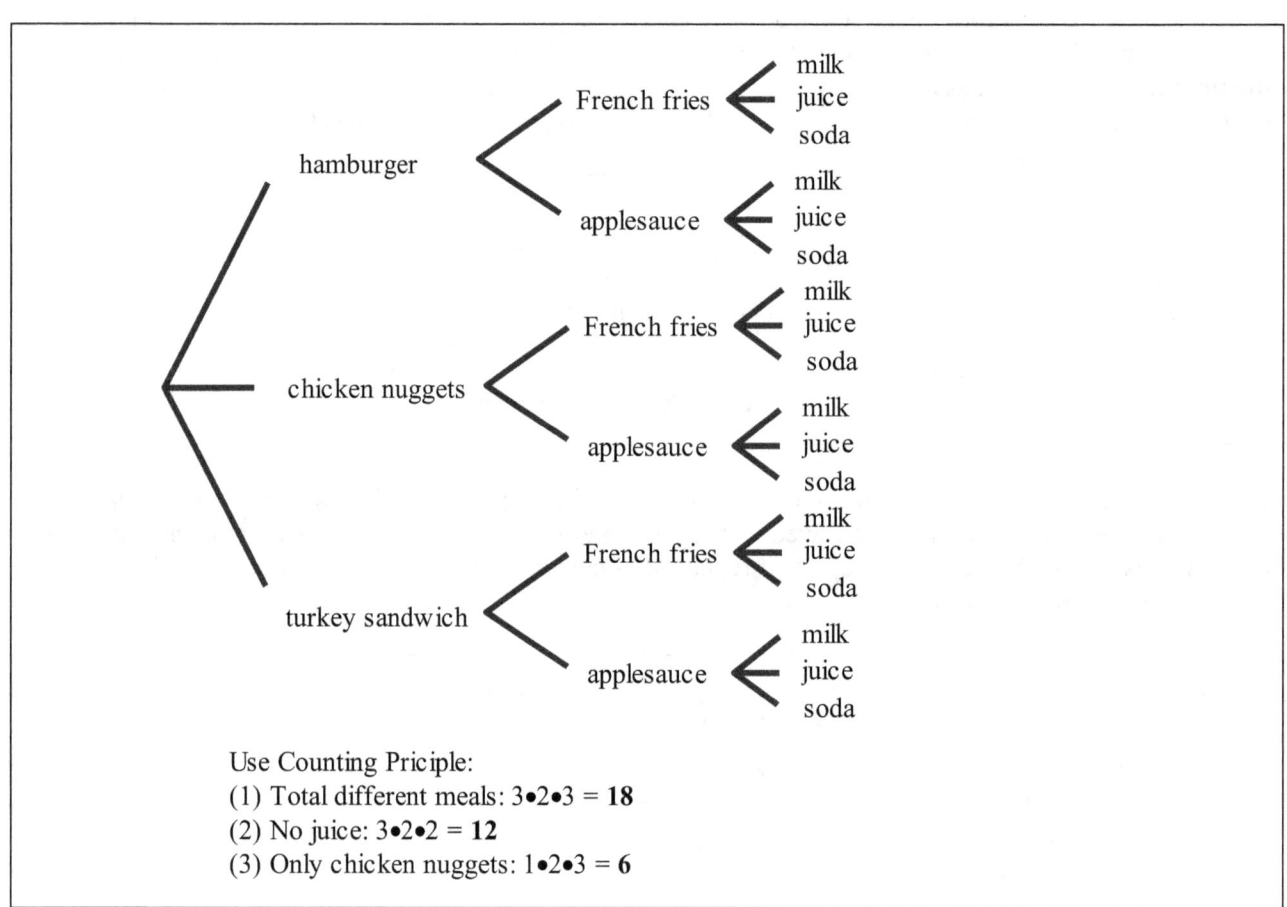

Use Counting Priciple:
(1) Total different meals: 3•2•3 = **18**
(2) No juice: 3•2•2 = **12**
(3) Only chicken nuggets: 1•2•3 = **6**

1. A school wants to add a coed soccer program. To determine student interest in the program, a survey will be taken. In order to get an unbiased sample, which group should the school survey?
*(1) every third student entering the building
 (2) every member of the varsity football team
 (3) every member in Ms. Zimmer's drama classes
 (4) every student having a second-period French class

2. Which data set describes a situation that could be classified as qualitative?
 (1) the elevations of the five highest mountains in the world
 (2) the ages of presidents at the time of their inauguration
*(3) the opinions of students regarding school lunches
 (4) the shoe sizes of players on the basketball team

The qualitative data is not numerical.

---

3. Alex earned scores of 60, 74, 82, 87, 87, and 94 on his first six algebra tests. What is the relationship between the measures of central tendency of these scores?
 (1) median < mode < mean   (3) mode < median < mean
 (2) mean < mode < median   ***(4) mean < median < mode**

$$\text{mean} = \frac{60 + 74 + 82 + 87 + 87 + 94}{6} = 80.7$$
$$\text{median} = \frac{82 + 87}{2} = 84.5$$
$$\text{mode} = 87$$

---

4. A survey is being conducted to determine which types of television programs people watch. Which survey and location combination would likely contain the most bias?
*(1) surveying 10 people who work in a sporting goods store   (3) randomly surveying 50 people during the day in a mall
 (2) surveying the first 25 people who enter a grocery store   (4) randomly surveying 75 people during the day in a clothing store

---

The sample must be large enough to be effective and must be chosen randomly to eliminate any bias. 10 people is not large enough and people who work in a sporting goods store is a sepecial group.

---

5. Which data set describes a situation that could be classified as qualitative?
 (1) the ages of the students in Ms. Marshall's Spanish class   ***(3) the favorite ice cream flavor of each of Mr. Hayden's students**
 (2) the test scores of the students in Ms. Fitzgerald's class   (4) the heights of the players on the East High School basketball team

---

The qualitative data is not numerical.

6. The freshman class held a canned food drive for 12 weeks. The results are summarized in the table below.

Canned Food Drive Results

Week	1	2	3	4	5	6	7	8	9	10	11	12
Number of Cans	20	35	32	45	58	46	28	23	31	79	65	62

Which number represents the second quartile of the number of cans of food collected?

(1)   29.5   (2)   30.5   ***(3)**   **40**   (4)   60

First rearrange the data in numerical order:
20, 23, 28, 31, 31, 35, 45, 46, 58, 62, 65, 79   (make sure 12 numbers)

the second quartile is the median $= \dfrac{35 + 45}{2} = 40$

---

7. Which data table represents univariate data?

(1)

Side Length of a Square	Area of Square
2	4
3	9
4	16
5	25

(2)

Hours Worked	Pay
20	$160
25	$200
30	$240
35	$280

*(3)

Age Group	Frequency
20–29	9
30–39	7
40–49	10
50–59	4

(4)

People	Number of Fingers
2	20
3	30
4	40
5	50

Frequency is the number of times that a value occurs, it does not represent another variate.

---

8. Which table does *not* show bivariate data?

(1)

Height (inches)	Weight (pounds)
39	50
48	70
60	90

*(3)

Quiz Average	Frequency
70	12
80	15
90	6

(2)

Gallons	Miles Driven
15	300
20	400
25	500

(4)

Speed (mph)	Distance (miles)
40	80
50	120
55	150

> Frequency is not considered as a variate in bivariate data.

9. Four hundred licensed drivers participated in the math club's survey on driving habits. The table below shows the number of drivers surveyed in each age group.

Ages of People in Survey on
Driving Habits

Age Group	Number of Drivers
16–25	150
26–35	129
36–45	33
46–55	57
56–65	31

Which statement best describes a conclusion based on the data in the table?
(1) It may be biased because no one younger than 16 was surveyed.
(2) It would be fair because many different age groups were surveyed.
(3) It would be fair because the survey was conducted by the math club students.
*(4) **It may be biased because the majority of drivers surveyed were in the younger age intervals.**

10. The box-and-whisker plot below represents the math test scores of 20 students.

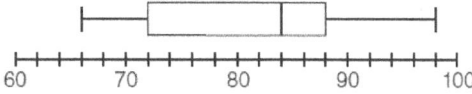

What percentage of the test scores are *less than* 72?
*(1)   25   (2)   50   (3)   75   (4)   100

11. The data set 5, 6, 7, 8, 9, 9, 9, 10, 12, 14, 17, 17, 18, 19, 19 represents the number of hours spent on the Internet in a week by students in a mathematics class. Which box-and-whisker plot represents the data?

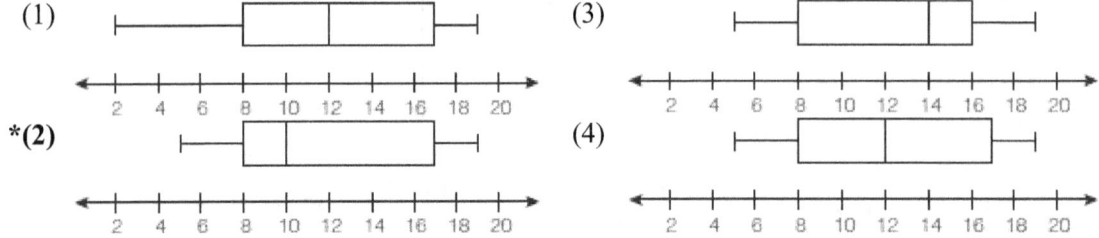

(1)

(3)

*(2)

(4)

> The 8th number is the median: 10.

12. The box-and-whisker plot below represents students' scores on a recent English test.

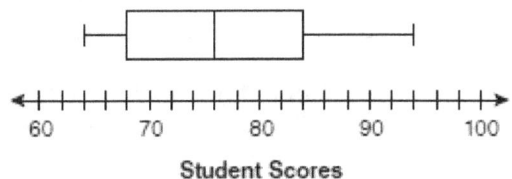

Student Scores

What is the value of the upper quartile?
(1)   68   (2)   76   *(3)   84   (4)   94

13. A movie theater recorded the number of tickets sold daily for a popular movie during the month of June. The box-and-whisker plot shown below represents the data for the number of tickets sold, in hundreds.

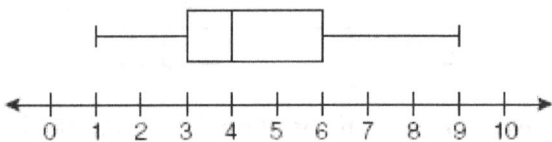

Which conclusion can be made using this plot?
(1) The second quartile is 600.   (3) The range of the attendance is 300 to 600.
(2) The mean of the attendance is 400.   *(4) **Twenty-five percent of the attendance is between 300 and 400.**

> (1) The second quartile is 400, not 600.
> (2) The mean can not be read on the box-and-whisker plot.
> (3) The range of the attendance is 100 to 900, not 300 to 600.
> (4) 300 is the first quartile; 400 is the second quartile.
>   25% of the data values are between the first quartile and the second quartile.

14. The table below shows a cumulative frequency distribution of runners' ages.

Cumulative Frequency Distribution of Runners' Ages

Age Group	Total
20–29	8
20–39	18
20–49	25
20–59	31
20–69	35

According to the table, how many runners are in their forties?

(1)   25   (2)   10   ***(3)   7**   (4)   6

> Runners in age group 20 - 49 is 25.
> Runners in age group 20 - 39 is 18.
> Runners in their forties is 25 - 18 = **7**

15. Which situation describes a correlation that is *not* a causal relationship?
*(1) The rooster crows, and the Sun rises.
 (2) The more miles driven, the more gasoline needed
 (3) The more powerful the microwave, the faster the food cooks.
 (4) The faster the pace of a runner, the quicker the runner finishes.

> Causation is the relationship in which one variable produces an effect on the other.

16. Which situation should be analyzed using bivariate data?
 (1) Ms. Saleem keeps a list of the amount of time her daughter spends on her social studies homework.
*(2) Mr. Benjamin tries to see if his students' shoe sizes are directly related to their heights.
 (3) Mr. DeStefan records his customers' best video game scores during the summer.
 (4) Mr. Chan keeps track of his daughter's algebra grades for the quarter.

17. Which situation describes a correlation that is *not* a causal relationship?
 (1) the length of the edge of a cube and the volume of the cube
 (2) the distance traveled and the time spent driving
*(3) the age of a child and the number of siblings the child has
 (4) the number of classes taught in a school and the number of teachers employed

18. There is a negative correlation between the number of hours a student watches television and his or her social studies test score. Which scatter plot below displays this correlation?

(1)

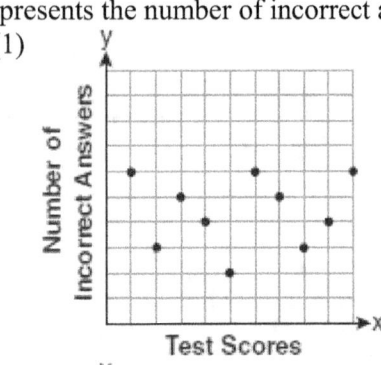

(3)

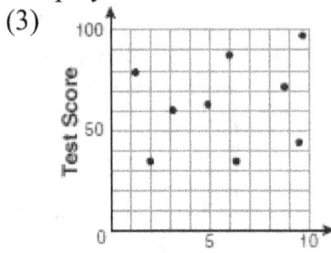

(2)

*(4)

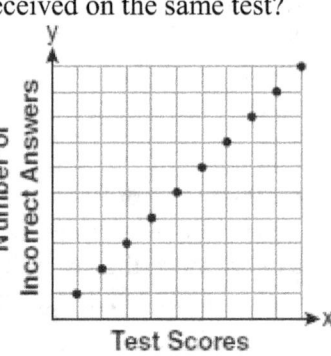

19. Which scatter plot shows the relationship between *x* and *y* if *x* represents a student score on a test and *y* represents the number of incorrect answers a student received on the same test?

(1)

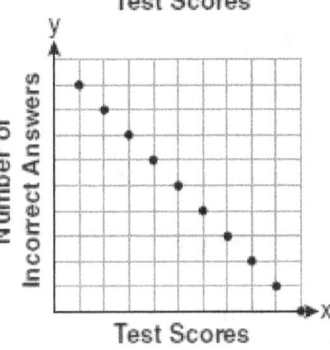

(3)

(4)

*(2)

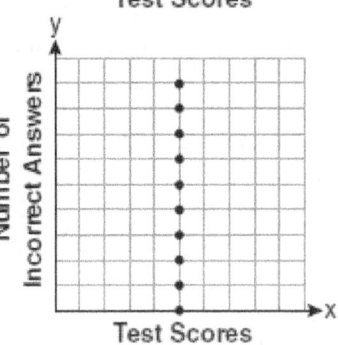

Show Work:

1. The values of 11 houses on Washington St. are shown in the table below.

Value per House	Number of Houses
$100,000	1
$175,000	5
$200,000	4
$700,000	1

Find the mean value of these houses in dollars. Find the median value of these houses in dollars. State which measure of central tendency, the mean or the median, *best* represents the values of these 11 houses. Justify your answer.

$$\text{mean} = \frac{100,00 + 175,000 \cdot 5 + 200,000 \cdot 4 + 700,000}{11} = 225,000$$

median = 175,000   (the value of the 6th number)

700,000 is far outside most of the values in the data set. When outliers exist, it is better to use the median to represent the values of these 11 houses.

2. Ms. Mosher recorded the math test scores of six students in the table below.

Student	Student Score
Andrew	72
John	80
George	85
Amber	93
Betty	78
Roberto	80

Determine the mean of the student scores, to the *nearest tenth*. Determine the median of the student scores. Describe the effect on the mean and the median if Ms. Mosher adds 5 bonus points to each of the six students' scores.

Rearrange the data in numerical order: 72, 78, 80, 80, 85, 93

$$\text{mean} = \frac{72 + 78 + 80 + 80 + 85 + 93}{6} = 81.3, \qquad \text{median} = \frac{80 + 80}{2} = 80$$

After 5 points added, both mean and median are increased by 5 points.
mean = 81.3 + 5 = 86.3 ,   median = 80 + 5 = 85

3. The Fahrenheit temperature readings on 30 April mornings in Stormville, New York, are shown below.

41°, 58°, 61°, 54°, 49°, 46°, 52°, 58°, 67°, 43°, 47°, 60°, 52°, 58°, 48°,
44°, 59°, 66°, 62°, 55°, 44°, 49°, 62°, 61°, 59°, 54°, 57°, 58°, 63°, 60°

Using the data, complete the frequency table below.

Interval	Tally	Frequency
40 - 44	////	4
45 - 49	ЖЖ	5
50 - 54	////	4
55 - 59	ЖЖ ///	8
60 - 64	ЖЖ //	7
65 - 69	//	2

On the grid below, construct and label a frequency histogram based on the table.

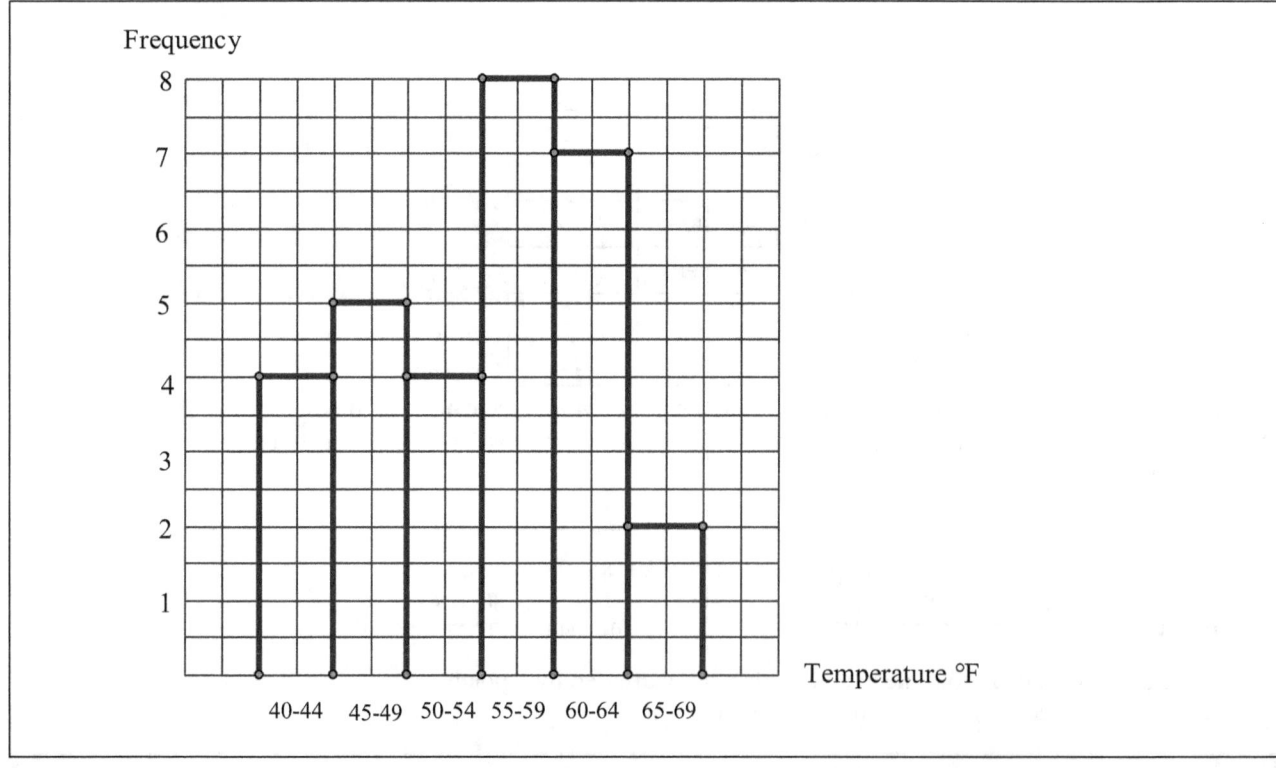

4. The diagram below shows a cumulative frequency histogram of the students' test scores in Ms. Wedow's algebra class.

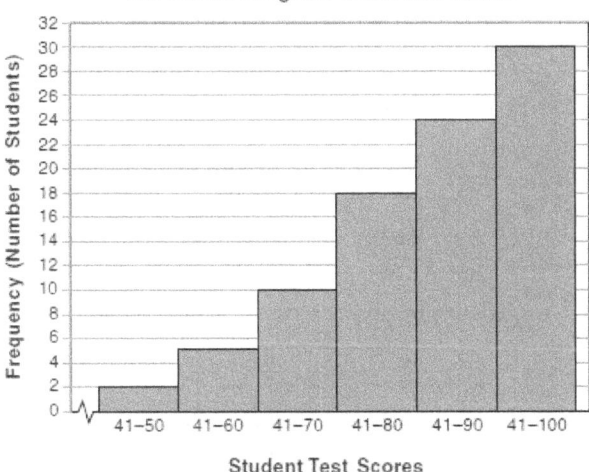

Ms. Wedow's Algebra Class Test Scores

Determine the total number of students in the class. Determine how many students scored higher than 70. State which *ten-point interval* contains the median. State which *two ten-point* intervals contain the same frequency.

---

    (1) The total number of students in the class is 30.
    (2) The number of students scored higher than 70 is 30 - 10 = 20.
    (3) The interval 71 - 80 contains the median. (The two middle numbers are the 15th and the 16th.)
    (4) The number of students scored in the interval 91 - 100 is 30 - 24 = 6.
        The number of students scored in the interval 81 - 90 is 24 - 18 = 6.
        Therefore the interval 81 - 90 and the interval 91 - 100 have the same frequency 6.

5. Megan and Bryce opened a new store called the Donut Pit. Their goal is to reach a profit of $20,000 in their 18th month of business. The table and scatter plot below represent the profit, P, in thousands of dollars, that they made during the first 12 months.

t (months)	P (profit, in thousands of dollars)
1	3.0
2	2.5
3	4.0
4	5.0
5	6.5
6	5.5
7	7.0
8	6.0
9	7.5
10	7.0
11	9.0
12	9.5

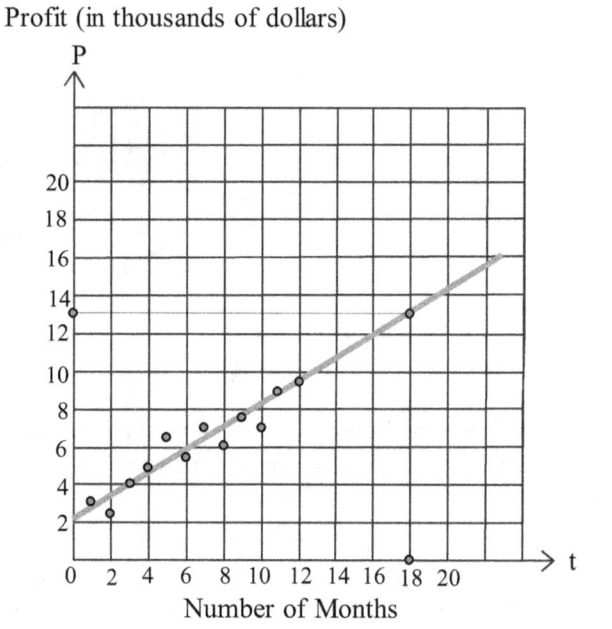

Profit (in thousands of dollars)

Number of Months

Draw a reasonable line of best fit. Using the line of best fit, predict whether Megan and Bryce will reach their goal in the 18th month of their business. Justify your answer.

Draw a reasonable line of best fit among these points.
According to the line of best fit, they can only reach $13,000 in the 18th month.
They can not reach the goal of $20,000.

6. The table below shows the number of prom tickets sold over a ten-day period.

Prom Ticket Sales

Day (x)	1	2	5	7	10
Number of Prom Tickets Sold (y)	30	35	55	60	70

Plot these data points on the coordinate grid below. Use a consistent and appropriate scale. Draw a reasonable line of best fit and write its equation.

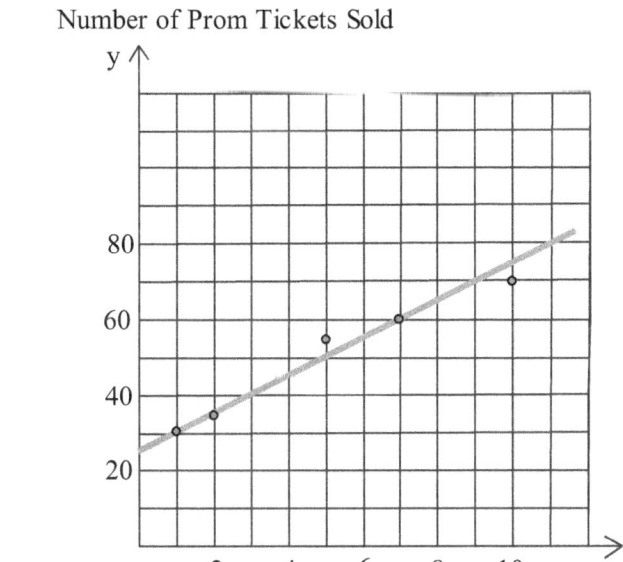

Draw a reasonable line among these points.

Choose two points on the line: (1, 30) and (7, 60).

$$m = \frac{y_2 - y_1}{x_2 - x_1} = \frac{60 - 30}{7 - 1} = \frac{30}{6} = 5$$

(note: Do not count the number of the grids to find the slope, since they use different scales.)

$b = 25$

$y = 5x + 25$

www.ingramcontent.com/pod-product-compliance
Lightning Source LLC
Chambersburg PA
CBHW081136170526
45165CB00008B/2689